Socio-Affective Computing

Volume 8

Series Editors
Amir Hussain, University of Stirling, Stirling, UK
Erik Cambria, Nanyang Technological University, Singapore

This exciting Book Series aims to publish state-of-the-art research on socially intelligent, affective and multimodal human-machine interaction and systems. It will emphasize the role of affect in social interactions and the humanistic side of affective computing by promoting publications at the cross-roads between engineering and human sciences (including biological, social and cultural aspects of human life). Three broad domains of social and affective computing will be covered by the book series: (1) social computing, (2) affective computing, and (3) interplay of the first two domains (for example, augmenting social interaction through affective computing). Examples of the first domain will include but not limited to: all types of social interactions that contribute to the meaning, interest and richness of our daily life, for example, information produced by a group of people used to provide or enhance the functioning of a system. Examples of the second domain will include, but not limited to: computational and psychological models of emotions, bodily manifestations of affect (facial expressions, posture, behavior, physiology), and affective interfaces and applications (dialogue systems, games, learning etc.). This series will publish works of the highest quality that advance the understanding and practical application of social and affective computing techniques. Research monographs, introductory and advanced level textbooks, volume editions and proceedings will be considered.

More information about this series at http://www.springer.com/series/13199

Soujanya Poria • Amir Hussain • Erik Cambria

Multimodal Sentiment Analysis

Springer

Soujanya Poria
Artificial Intelligence Initiative
ASTAR
Singapore, Singapore

Erik Cambria
School of Computer Engineering
Nanyang Technological University
Singapore, Singapore

Amir Hussain
Computing Science and Mathematics
University of Stirling
Stirling, UK

ISSN 2509-5706 ISSN 2509-5714 (electronic)
Socio-Affective Computing
ISBN 978-3-030-06956-8 ISBN 978-3-319-95020-4 (eBook)
https://doi.org/10.1007/978-3-319-95020-4

This Springer imprint is published by the registered company Springer Nature Switzerland AG
The registered company address is: Gewerbestrasse 11, 6330 Cham, Switzerland

Preface

Emotions and sentiments play a crucial role in our everyday lives. They aid decision-making, learning, communication, and situation awareness in human-centric environments. Over the past two decades, researchers in artificial intelligence have been attempting to endow machines with cognitive capabilities to recognize, infer, interpret, and express emotions and sentiments. All such efforts can be attributed to affective computing, an interdisciplinary field spanning computer science, psychology, social sciences, and cognitive science. Sentiment analysis and emotion recognition has also become a new trend in social media, avidly helping users understand opinions being expressed on different platforms in the web. In this book, we focus on developing novel methods for text-based sentiment analysis. As an application of the developed methods, we employ them to improve multimodal polarity detection and emotion recognition. Specifically, we develop innovative text- and visual-based sentiment-analysis engines and use them to improve the performance of multimodal sentiment analysis. We begin by discussing challenges involved in both text-based and multimodal sentiment analysis. Next, we present a number of novel techniques to address these challenges. In particular, in the context of concept-based sentiment analysis, a paradigm gaining increasing interest recently, it is important to identify concepts in text; accordingly, we design a syntax-based concept-extraction engine. We then exploit the extracted concepts to develop concept-based affective vector space which we term, "EmoSenticSpace." We then use this for deep learning-based sentiment analysis, in combination with our novel linguistic pattern-based affective reasoning method termed *sentiment flow*. Finally, we integrate all our text-based techniques and combine them with a novel deep learning-based visual feature extractor for multimodal sentiment analysis and emotion recognition. Comparative experimental results using a range of benchmark datasets have demonstrated the effectiveness of the proposed approach.

Singapore, Singapore — Soujanya Poria
Stirling, UK — Amir Hussain
Singapore, Singapore — Erik Cambria

Contents

Chapter 1
Introduction and Motivation

Abstract Multimodal sentiment analysis a new research field in the area of Artificial Intelligence. It aims at processing multimodal inputs for e.g., Audio, Visual and Text to extract affective knowledge. In this chapter we discuss the major research challenges in this topic followed by the overview of the proposed multimodal sentiment analysis framework.

Keywords Multimodal sentiment analysis · Emotion recognition · Emosenticnet · Sentic patterns · Deep learning

People increasingly tend to share online their opinions and experiences regarding products or services using forums, blogs, social networks, chats and content-sharing services. Such opinions strongly influence other people's decisions to purchase a product, watch a movie, visit a resort, vote for a political party and so on. For example, in a social network such personal opinions have been shown to have stronger impact on users than paid advertising [31].

The tremendous amount of available data requires its automatic analysis in order for these data to be useful in decision-making. Examples of such automatic analysis include identifying the popularity of movies from online reviews, determining which model of a camera has the best value or which piece of music most teenagers like. Such analysis lets consumers learn about advantages and disadvantages of a product or service from experiences shared by other users. It also lets providers and producers extract important information from the feedback from individual users, as well as from trends in opinion or sentiment across specific populations or regions, to help drive continued improvement.

Automatic analysis of user-contributed opinions improves the income of business by allowing them to adapt their products and services to users' needs and thus increase sales. It improves efficiency of political parties by allowing them to adapt their actions to public opinion and thus attract more votes. More importantly, it improves the quality of life of all of us, users, consumers and citizens, both those who actively use social media and even those who don't, by providing us with improved products and services offered by businesses and with improved actions

S. Poria et al., *Multimodal Sentiment Analysis*, Socio-Affective Computing 8,
https://doi.org/10.1007/978-3-319-95020-4_1

of local administration and state government. Another way in which it significantly improves our quality of life is allowing us make wiser and more informed buying decisions, thus avoiding bad products and services and choosing the best ones. Finally, in political context it enables democracy in real time: instead of voting for a new government once in several years without even explaining why we vote for one party and against another one, we can now influence the policies and actions of the local administration and the state government by directly expressing our opinions in social media, which they monitor to adjust their policies.

Till recently user-contributed content has primarily been textual. However, a growing number of opinions are being posted in video format. Nowadays every smartphone is equipped with a high-definition camera, so it has become almost effortless for anyone to publish their opinion as a tweet, a picture of their meal on Instagram or a video on YouTube. It has become common for consumers to record their opinions of products in front of their webcams and upload them on social media such as YouTube or Facebook. These services allow their users to stream their experiences live, to friends and fans around the world. Users can easily praise a restaurant for a wonderful meal or can just as easily take footage of terrible customer service. Many users share comparison videos, showing products from competing brands side by side, with the pros and cons of the product discussed. Such information from ordinary users is typically unbiased and is therefore highly useful for people seeking to choose a product. Many terabytes of user-contributed video are now being published every day, and with an ever-growing number of affordable smartphones and video cameras, this will only grow.

Though video is costly to store and to analyse, it provides more information than text does. Indeed, humans communicate by expressing their emotions and sentiments When analysing text, the program can only deal with words, phrases, relationships and the dependencies among them. This is often not sufficient to understand affective content. More cues are available in video, with its audio and visual modalities, in terms of vocal modulations and facial expressions. This data can be used alongside textual data to provide important additional information that helps to identify affective state of the person. A combination of text and video data gives a better emotion and sentiment analysis model than either could alone. Human beings also rely more on multimodal information than unimodal [355]: we better understand a speaker's intention when we can see his or her facial expressions while he or she is speaking, because together aural and visual channels provide more information than they provide alone. This is a particular case of the general rule that the brain relies on several sources of sensory input to compensates for incomplete information of each channel separately; see Fig. 1.1.

A computing discipline that aims to automatic identification of private states of human mind, such as opinions, emotions, sentiments, behaviour and beliefs, from textual, audio, video or other data or their combination is called sentiment analysis. This discipline is part of affective computing, an emerging interdisciplinary field spanning artificial intelligence (AI), cognitive sciences, computer science and social sciences. Sentiment analysis can be performed at different levels. For example, subjectivity detection only classifies data into subjective (those that contain affective

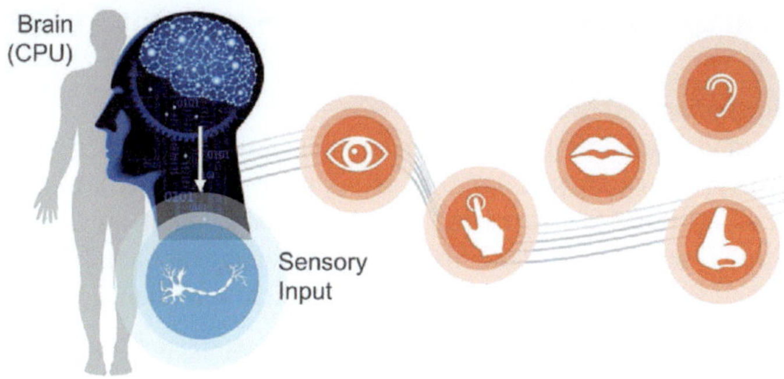

Fig. 1.1 Human brain considers multimodal information cumulatively for decision making

information) and objective. Polarity detection focuses on determination whether subjective data indicate positive or negative sentiment. Emotion recognition further determines a specific emotion expressed in the data: positive sentiment can be caused by joy or surprise, while negative sentiment can be caused by fear or disgust. Even though the primary focus of this work is to classify sentiment of the videos, we also show the performance of the proposed method in emotion recognition.

The interest in sentiment and emotion analysis research, both text-based and multimodal, has been greatly increasing in recent years, due to the wide range of business and social applications in human-centric environments it can apply to, such as education and healthcare. Making appropriate sense of what other people think has always been considered important for decision making. There exists a significant body of research on determining the sentiment or polarity of natural language text. Supervised, unsupervised and semi-supervised methods have made good progress when applied to text-based sentiment analysis. On benchmark datasets these methods are found to produce impressive accuracy. Although the use of audio and visual cues together for emotion detection is not as well studied, recent advances in computer vision have enabled researchers to obtain quite impressive results. Merging audio, visual and textual data for multimodal sentiment analysis is even less studied and so the effectiveness of this approach remains largely underestimated in the literature. Only a few research articles focusing on this topic are available in the literature [41, 239] and none of these have clearly described both the extraction of features and the fusion of the information extracted from different modalities.

In this book, we propose novel concept-level text analysis frameworks for sentiment analysis based on complex linguistic patterns and machine-learning techniques. As an application of the developed techniques, we show that these text-analysis frameworks are useful in multimodal sentiment analysis and emotion recognition, where they enhance the performance of a next-generation audio-visual sentiment analysis framework. To this end, we also propose a novel visual sentiment-analysis framework based on deep machine learning.

1.1 Research Challenges in Text-Based Sentiment Analysis

Our experiments have shown that text provides the most accurate affective clues of the three modalities considered separately, hence the importance of text-based sentiment analysis. In addition, currently text is the more readily available type of data, easier to store and to analyse and richer in information it conveys. However, in spite of a rich seventy-year-long history of automatic text analysis technique, a number of challenges in language processing applied to sentiment analysis remain.

Complex sentence structure For sentiment classification, it is crucial to understand the text structure. Consider, for example, the sentence "I do not like the movie that you love". Both positive and negative sentiment towards the movie are expressed in this sentence, and most systems will be unable to distinguish their impact on the persons's sentiments expressed in this sentence. In another example, "This movie doesn't care about cleverness, wit or any other kind of intelligent humor", although the sentence contains three words having positive polarity (*cleverness*, *wit*, *intelligent*), the overall sentiment is in fact negative: negation (*does not*) serves to flip the polarity of all these words.

The curse of dimensionality The majority of existing text classification systems, including commercial ones, use bag-of-words (BOW) and n-gram methodologies, using term frequency of the n-grams or words as features. Despite being simple, these methods have been found to be extremely effective in text classification. However, feature vectors generated by these methods often tend to be very sparse and high-dimensional. Such feature vectors have processing time infeasible at large scale and can cause overfitting even in the presence of regularization.

Implicit sentiment and desirable facts Explicit sentiment expressions are those in which specific sentiment words are used to express the opinions, such as *excellent* or *ugly*. Most often, however, people do not express their sentiment explicitly. Consider the sentence "I went to eat, but the shop was closed." Although there is no sentiment word marker present, this sentence still expresses negative polarity, because it communicates an undesirable fact. Identifying implicit sentiment is one of the most difficult challenges in the field of sentiment analysis.

Noisy data Data from Internet sources are most commonly unstructured and noisy: for instance, they may contain HTML tags or pieces of programs in scripting languages, as well as use informal and colloquial expressions. However, the types of noise are not limited to these; ungrammatical text, slang and micro text such as tweets, abbreviated phrases and words (e.g., "OMG", "LOL") all act as noise in the corpus. Unless these issues are addressed, an automated sentiment analysis framework does not yield good results.

Multiple topics in the review A review can contain multiple topics. A person reviewing a product can mention some another product at the same time, then abruptly change topic and discuss something else, etc. Reviews also address different features of a product, commonly referred to as aspects of the product. In "The battery of Galaxy S6 performs really well", *battery* is the aspect of the product *Galaxy S6* and the sentiment on this aspect expressed in this sentence is positive. In "Galaxy S6 is way too expensive" the negative opinion is expressed about the aspect *price*, which is not explicitly mentioned in the sentence (phenomenon known as implicit aspect). Identifying these aspects and their associated sentiment (as well as aggregating such sentiment from a large number of reviews) is a highly demanding task, known as aspect-based sentiment analysis.

Sarcasm, irony and metaphors Sarcasm detection is important for many natural language processing tasks. In sentiment analysis, sarcasm can flip the polarity of an apparently positive sentence and, hence, negatively affect polarity detection performance. Understanding the ground truth or the facts about a given event allows for detection of contradiction between the objective polarity of the event (usually negative) and its sarcastic characteristic by the author (usually positive), as in "I love the pain of breakup". Obtaining such knowledge is, however, very difficult. This makes sarcasm detection very important yet very challenging. Irony and metaphors are examples of other semantic challenges in text understanding relevant for sentiment analysis.

In this book, we address these challenges, except the last item, addressing which with the help of the developed tools is on the agenda for our future work.

1.2 Research Challenges in Multimodal Sentiment Analysis

Audio data in a video expresses the tone of the speaker and visual data conveys the facial expressions and movement, which in turn helps understanding the affective state of users. Video data is also an important source for determining emotion and sentiment analysis. However, Multimodal analysis is far more complex than unimodal (e.g. text-only), since multiple modalities are involved in the processing and research. Accordingly, major challenges remain in analysing audio and video data.

High-dimensional Feature Space As multiple modalities are involved, the feature space consisted of the feature vectors of all modalities are often very high. This is a serious concern in designing a multimodal sentiment analysis framework to perform in real-time. To tackle this problem, late fusion or decision-level fusion techniques have been proposed in literature, although they do not deliver good performance.

Noisy data channel A multimodal framework with noisy or incorrect data from one or more than one modality can greatly degrade performance. Therefore, a primary step in this research is to identify and discard those noisy mediums are rudimentary processes. For example, when a face is blurred due to poor video quality, audio and text modalities can then be prioritized to make decisions and vice-versa.

Combining information from multiple modalities The information extracted from all modalities should be combined together using any fusion technique for multimodal sentiment analysis. The challenges in the fusion process include determining which modality should contribute most in the fusion, filtering the noisy modality, and handling the high dimensional feature space in feature level fusion.

Individual expression style Expression of opinions varies from person to person [239]: one person, for example, may primarily express his or her opinions vocally, while another person may express himself or herself more visually. When a person expresses his or her opinions with increased vocal modulation, the audio data stream will likely contain most of the clues for sentiment analysis and opinion mining, for that particular individual. However, another person may be more communicative through facial expression and so for this person, the best data for mining sentiment, emotions and opinions would likely be found in the facial analysis. As a result, a generic model needs to be developed that can adapt itself for any user and still provide a consistent result.

In this book, we address, among other, these challenges. In particular, our multimodal affective content classification model has been trained on robust data containing opinions of many users and thus is user-independent.

1.3 Overview of the Proposed Framework

The proposed sentiment analysis methodology includes enhancements for both text-based sentiment analysis and multimodal sentiment analysis. In both cases, we perform sentence-level analysis, i.e., the input for our method is either a single sentence or a video clip containing a single utterance. Figure 1.2 shows the overall framework of the methodology developed in this book, which uses several text-processing modules (discussed in Chaps. 4, 5, and 6) as well as audio and video-processing modules (discussed in Chap. 7) for multimodal polarity detection and emotion recognition (in this section we describe the case of polarity detection; emotion recognition was done in a similar manner).

1.3.1 Text-Based Sentiment Analysis

As the primary method for text-based analysis tasks, we focus on concept-level analysis. Concepts are one-word or multi-word semantic elements, such as *holiday*, *cold beer* or *small screen*. Every concept-level text-analysis method requires a concept extraction step. As the first step, we extract the concepts using a novel dependency pattern-based concept-extraction method based on the syntactic structure of the sentences.

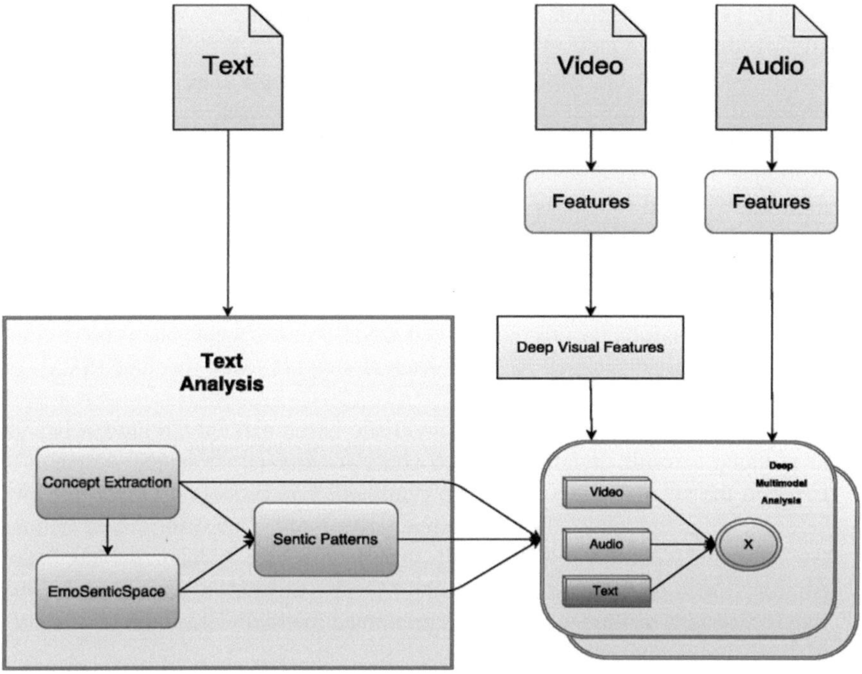

Fig. 1.2 Overall framework of the proposed method

After the concept extraction step, we apply one of the three alternative methods to obtain the polarity value for the sentence: a linguistic-based method, a supervised machine learning-based method, and their combination.

Linguistic-based method First, we associate real-valued polarity with each concept extracted from the sentence (for which we have such data) by retrieving it from a state-of-the-art concept-level sentiment lexicon SenticNet [45]. Then the obtained values are corrected and aggregated into a single value using a novel framework that we called sentic patterns. This set of linguistic patterns implement a novel methodology that we called sentiment flow: hierarchically assembling the polarity value by promoting the values of individual nodes (obtained from SenticNet) up the dependency syntax tree. The values can be inverted, increased or decreased in this process as they pass by a valence shifter such as negation or a logical connector such as "but" or "though". This assembling process results in one polarity value being associated with the root of the tree, which is used as the final classification result.

Supervised machine learning-based method First, we project the extracted concepts into a vector space. To this end, we significantly improved the method by Cambria et al. [43] using a novel automatic emotion lexicon generation technique based on fuzzy logic and support vector machine (SVM), which produced a concept-level lexicon that is larger in size and quality than the AffectNet resource

used in [43]. This lexicon, which we called EmoSenticNet, was subsequently employed to generate the affective concept vector space EmoSenticSpace, from which we retrieve the feature vectors for the concepts. The obtained feature vectors are aggregated into one feature vector of the sentence.
Independently, another feature vector is built by the sentence using a novel technique based on convolutional neural network (CNN). As input for this network we used a concatenation of pre-trained `word2vec` vectors [231], a state-of-the-art distributional semantics model, and as the resulting feature vector we used the activation values of the penultimate fully-connected layer of the trained CNN. Finally, the two sentence vectors—the one obtained via EmoSenticSpace and the one obtained via CNN—were concatenated and fed into a state-of-the-art supervised classifier such as support vector machine (SVM) or extreme learning machine (ELM).

Combined method The unsupervised linguistic-based method not always is able to produce a result: it fails when no concept extracted from the sentence is found in the polarity lexicon or when sentiment flow process does not find any values relevant to produce the final output. Given that the linguistic-based method typically produces more reliable results than the machine-learning method, in the combined method we used the machine-learning method as a back-off for the linguistic-based method. Overall, this combined method yielded the best results of the three methods.

1.3.2 Multimodal Sentiment Analysis

As an application of these concept-level text-analysis methods, we employed them to enhance multimodal sentiment analysis. In this case one can also consider three alternative methods: a feature-level fusion method, a decision-level fusion method, and a combined method.

Feature-level fusion method Given a video clip containing a single utterance, with text previously transcribed, we extracted features from textual, audio and visual modalities. Textual features were extracted using techniques explained in Sect. 1.3.1. To extract audio features, we used the existing state-of-the-art Opensmile software. To extract visual features, we used two alternative methods. In one of them, we recognized the face in the image and employed existing state-of-the-art facial feature extraction tool, CLM-Z. In another, enhanced method, we extracted the features directly from the image using a novel technique we developed, which combines a deep convolutional neural network (CNN) with time-delayed recurrent neural network (RNN); we called the constructed type of network convolutional recurrent neural network (CRNN). Then the obtained textual, audio, and visual features were concatenated into a large feature vector of the clip. These feature vectors were used to train a state-of-the-art supervised classifier.

Decision-level fusion method In this case, a classifier for each modality is trained separately. For only textual modality, any of the classifiers described in Sect. 1.3.1 can be used. For only audio and video modalities, a state-of-the-art supervised classifier is applied to the features described above. Then, the obtained values for the three modalities are combined into a new, short feature vector of the clip, and a state-of-the-art supervised classifier is trained on such vectors.

However, we did not apply pure decision-level fusion method in this book, because it gave poor results in another our work [290]. Instead, we used a more sophisticated combined fusion method described below.

Combined feature/decision-level fusion method In this case, a feature-level fusion classifier was trained for two modalities, and a separate classifier was used for the remaining modality (we tried all three possible pairs). The The process was as described in the two previous cases, correspondingly.

The output of both classifiers (the two-modal feature-level fusion one and the unimodal one) was interpreted as real-valued probabilities of each category; in our case this gave four values: two classifier responses for two categories each, positive and negative polarity. Then, decision-level fusion was applied by training a state-of-the-art supervised classifier on the set of these 4-dimensional vectors. As a variant, a fifth coordinate was added obtained from the linguistic-based text-only classifier described in Sect. 1.3.1.

1.4 Contributions of This Book

In this book, we address text-based and multimodal sentiment analysis tasks such as polarity detection and emotion recognition. The contributions of this book can be subdivided into two areas: novel text-based sentiment analysis frameworks and combining these text-based methods with multimodal techniques to improve performance of audio-visual sentiment analysis. The major contributions of this book can be summarized as follows.

1.4.1 Text-Based Sentiment Analysis

In our experience, text provides the most reliable sentiment analysis clues of the three modalities. In addition, currently textual data are more abundant in Internet than multimodal data.

Concept extraction techniques We have designed a novel concept-extraction algorithm based on linguistic patterns, which is able to extract concepts from text by exploiting the syntactic structure. This concept-extraction paradigm is the primary step for any concept-level text-analysis task. Comparison of our proposed concept-extraction method with the state-of-the-art concept-extraction method based on part-of-speech (POS) tags has shown that our model outperforms the state of the art by a 6% margin.

Technique for construction of concept-level emotion lexicon We have shown how fuzzy clustering can be used together with support vector machine (SVM) clasifier for enhanced classification. We have applied this technique to produce an emotion lexicon that we called EmoSenticNet. We have used this emotion lexicon to create a commonsense vector space of concepts that we called EmoSeticSpace, where each concept is represented as a 100-dimensional vector. We have also shown that this resource is very useful in several other text analysis tasks, such as personality detection.

Sentic patterns We have proposed a novel text-based sentiment-analysis method that fuses linguistic patterns with machine learning basing on sentence structure. We have developed a novel paradigm for real-time concept-level sentiment analysis that merges computational intelligence, linguistics, and commonsense computing in order to improve the accuracy of time-consuming tasks such as polarity detection in challenging ambiguous environments. Our analysis of the way sentiment flows from concept to concept through dependency relations helps the algorithm better understand the contextual role of each concept in text, and thus achieve a dynamic polarity inference that outperforms supervised state-of-the-art statistical methods in terms of accuracy.

CNN-based features for text analysis We have constructed a convolutional neural network (CNN) for automatic extraction of opinionated features from natural-language text. Specifically, we have developed a multilayer CNN with a varying filter size. These features have shown impressive performance in sentiment analysis. In comparison with manually engineered syntactic features such as POS tags or n-grams, our CNN-based textual features provide a notable performance improvement.

Combining sentic patterns with machine learning A major contribution of this book is to blend linguistic patterns with machine learning for sentiment analysis. This process ensures advantages of both methods, since when one method fails to detect sentiment, the other method provide the classification result. Extensive comparative evaluation shows that the joint exploitation of machine learning and linguistic patterns outperforms existing state-of-the-art approaches.

1.4.2 Multimodal Sentiment Analysis

We leverage the fact that multimodal data fusion increases the accuracy and reliability of estimates [309].

Distance-based facial features Once the faces are recognized in a video, how facial features can be extracted is well-studied in existing literature. We have addressed this issue by calculating several distance-based features of the facial characteristics points. These features have been shown to provide state-of-the-art performance on benchmark datasets. In particular, in the sentiment-analysis task these distances-based facial features have performed better than features represented by facial characteristic points directly.

Deep learning-based visual features We have designed convolutional recurrent neural network (CRNN), a novel architecture that combines a deep CNN with time-delayed recurrent neural network (RNN). Visual features automatically extracted with CRMM further boost the performance of our proposed multimodal sentiment analysis framework and outperform the state of the art.

Combining textual, visual and audio features for sentiment analysis We have developed a trimodal sentiment analysis framework by combining textual, visual and audio clues available in videos. Our experimental results have shown that trimodal analysis of opinionated videos produces better performance than unimodal and bimodal sentiment analysis. In particular, we have shown that deep network-based textual features play an important role in this process. With this, we have demonstrated the importance and feasibility of developing a multimodal framework that could cope with all three sensing modalities (text, audio, and video) in human-centric environments. In particular, we have shown that textual, audio, and visual modalities can be exploited together to enable effective extraction of semantic and affective information conveyed in communication. Specifically, we have shown that ensemble application of feature extraction from different types of data and modalities enhance the performance of the overall system.

1.5 Book Organisation

This book is organised as follows.

Chapter 2 provides theoretical details of the techniques and tools used in this research. First sentiment analysis and aspects are defined, followed by a review of several commonly used, state-of-the-art machine learning classifiers such as CNN, SVM and ELM, which were employed in this research. Evaluation methods for these machine-learning classifiers are also presented. Since word-embedding methods play an important role in our research, we extensively discuss a state-of-the-art word embedding algorithm (`word2vec`).

Chapter 3 reviews recent literature on text-based and multimodal sentiment analysis and emotion recognition. We specifically discuss the role of text analysis in multimodal sentiment analysis, including benchmark datasets and shortcomings of state of the art. We present quantitative evidence collected from literature to shows that multimodal sentiment analysis has been gaining major interest from the part of artificial intelligence researchers.

Chapter 4 presents a novel linguistic pattern-based approach for concept extraction from text. Concept extraction is the key step for any concept-level text-analysis framework. We identify linguistic patterns for extracting concepts from the dependency parse tree of natural language sentences. The algorithm is shown to perform very well on benchmark data, achieving 92.21% accuracy. We also show extensive applications of the concept parser in various natural language processing applications.

Chapter 5 describes the method of developing an affective reasoning knowledge base that we called EmoSenticSpace. In order to build EmoSenticSpace, we develop a concept-level emotion lexicon EmoSenticNet by using a fuzzy-based support vector machine (SVM) classifier. Next, we blend EmoSenticNet with the existing semantic resource ConceptNet, followed by singular-value decomposition (SVD) of the obtained matrix. We show that EmoSenticSpace is a particularly useful resource, which provides features that useful in different text-analysis tasks.

Chapter 6 introduces novel linguistic-based patterns for concept-level sentiment analysis. Upon obtaining the dependency trees of sentences, linguistic patterns are applied on them for sentiment analysis. In addition to introducing these linguistic patterns, we also propose the process of merging linguistic patterns with deep learning in order to further improve text-based sentiment analysis. Comparative experimental results showed that the combination of linguistic patterns and deep learning is very efficient for sentiment analysis.

Chapter 7 focuses on multimodal polarity detection and emotion recognition by integrating textual information extracted using the methods described in Chapters 4 to 6 with audio and visual features. We also discuss the development of a novel deep network for visual feature extraction. In addition, we evaluate and discuss in detail the impact of textual and visual features.

Chapter 8 concludes the book by summarizing the proposed methods, explaining the main benefits of this research and identifying its limitations. We also outline our ongoing work-in-progress that explores novel ways of fusion for different modalities by accounting for temporal dependencies of information within utterance presented in videos. Finally, we address the limitations of our current research by proposing a number of future research directions.

Chapter 2
Background

Abstract In this chapter we describe the theories and methods which are broadly utilized in this book. Starting with the notion of affective computing, sentiment analysis and opinion mining, this chapter covers the classifiers such as Support Vector Machine (SVM), Extreme Learning Machine (ELM), the basic model evaluation techniques e.g., precision, recall etc. We also explain Principal Component Analysis (PCA) and powerful textual features *Word Embeddings (Word2Vec)*.

Keywords Affective computing · Sentiment analysis · Senticnet · PCA · MSE

2.1 Affective Computing

Affective computing is the set of techniques aimed at performing affect recognition from data, in different modalities and at different granularity scales. Sentiment analysis, for example, performs coarse-grained affect recognition, as it is usually considered a binary classification task (positive versus negative), while emotion recognition performs fine-grained affect recognition, as it aims to classify data according to a large set of emotion labels.

While there is a fixed taxonomy for sentiment which is bound within positive, negative and neutral sentiments, the taxonomy for emotions is diverse. Philosophical studies on emotions date back to ancient Greeks and Romans. Following the early Stoics, for example, Cicero enumerated and organized the emotions into four basic categories: *metus* (fear), *aegritudo* (pain), *libido* (lust), and *laetitia* (pleasure). Studies on evolutionary theory of emotions, in turn, were initiated in the late nineteenth century by Darwin [78]. His thesis was that emotions evolved via natural selection and, therefore, have cross-culturally universal counterparts. In the early 1970s, Ekman found evidence that humans share six basic emotions: happiness, sadness, fear, anger, disgust, and surprise [104]. Few tentative efforts to detect non-basic affective states, such as fatigue, anxiety, satisfaction, confusion, or frustration, have been also made [77, 267, 307].

S. Poria et al., *Multimodal Sentiment Analysis*, Socio-Affective Computing 8,
https://doi.org/10.1007/978-3-319-95020-4_2

In 1980, Averill put forward the idea that emotions cannot be explained strictly on the basis of physiological or cognitive terms. Instead, he claimed that emotions are primarily social constructs; hence, a social level of analysis is necessary to truly understand the nature of emotions. The relationship between emotions and language (and the fact that the language of emotions is considered a vital part of the experience of emotions) has been used by social constructivists and anthropologists to question the universality of Ekman's studies, arguably because the language labels he used to code emotions are somewhat US-centric. In addition, other cultures might have labels that cannot be literally translated to English (e.g., some languages do not have a word for fear [329]).

For their deep connection with language and for the limitedness of the emotional labels used, all such categorical approaches usually fail to describe the complex range of emotions that can occur in daily communication. The dimensional approach [253], in turn, represents emotions as coordinates in a multi-dimensional space. For both theoretical and practical reasons, an increasing number of researchers prefer to define emotions according to two or more dimensions. An early example is Russell's circumplex model [328], which uses the dimensions of arousal and valence to plot 150 affective labels.

Similarly, Whissell considers emotions as a continuous 2D space whose dimensions are evaluation and activation [391]. The evaluation dimension measures how a human feels, from positive to negative. The activation dimension measures whether humans are more or less likely to take some action under the emotional state, from active to passive. In her study, Whissell assigns a pair of values <activation, evaluation> to each of the approximately 9,000 words with affective connotations that make up her Dictionary of Affect in Language.

Another bi-dimensional model is Plutchik's wheel of emotions, which offers an integrative theory based on evolutionary principles [278]. Following Darwin's thought, the functionalist approach to emotions holds that emotions have evolved for a particular function, such as to keep the subject safe [119]. Emotions are adaptive as they have a complexity born of a long evolutionary history and, although we conceive emotions as feeling states, Plutchik says the feeling state is part of a process involving both cognition and behavior and containing several feedback loops. In 1980, he created a wheel of emotions, which consisted of 8 basic emotions and 8 advanced emotions each composed of 2 basic ones. In such model, the vertical dimension represents intensity and the radial dimension represents degrees of similarity among emotions.

Besides bi-dimensional approaches, a commonly used framework for emotion representation is the <arousal, valence, dominance> set, which is known in the literature by different names, including <evaluation, activation, power> and <pleasure, arousal, dominance> [227]. Recent evidence suggests there should be a fourth dimension: Fontaine et al. reported consistent results from various cultures where a set of four dimensions is found in user studies, namely <valence, potency, arousal, unpredictability> [117]. Dimensional representationsof affect are attractive mainly because they provide a way of describing emotional states that is more tractable than using words. This is of particular importance when dealing with naturalistic data, where a wide range of emotional states occurs. Similarly, they are much more able to deal with non-discrete emotions and variations in emotional states over time [68], since in such cases changing from one universal emotion label to another would not make much sense in real life scenarios.

Dimensional approaches, however, have a few limitations. Although the dimensional space allows for comparing affect words according to their reciprocal distance, it usually does not allow making operations between these, e.g., for studying compound emotions. Most dimensional representations, moreover, do not model the fact that two or more emotions may be experienced at the same time. Eventually, all such approaches work at word level, which makes them unable to grasp the affective valence of multiple-word concepts.

All such limitations are overcome by the Hourglass of Emotions [44], a new affective representation based on Plutchik's model, which represents affective states both through labels and through four independent but concomitant affective dimensions, which can potentially describe the full range of emotional experiences that are rooted in any of us (Fig. 2.1). By leveraging on multiple (polarized) activation levels for each affective dimension, the Hourglass of Emotions covers cases where up to four emotions can be expressed at the same time and allows for reasoning on them in an algebraic manner. The model also allows for affective commonsense reasoning on both single words and multiple-word expressions [37] and provides a formula to calculate polarity based on emotions, which represents the first explicit attempt to bridge sentiment analysis and emotion recognition.

2.2 Sentiment Analysis

In order to take a decision of purchasing a product or watching a movie, for an individual it is nearly impossible to manually read thousands of opinions left by others available online. What if, Artificial Intelligence helps us to let us the overall public sentiment on the product we desire? Sentiment Analysis and Opinion Mining is that AI method which analyzes opinions and extract users' sentiment. Opinion mining can tell us whether a product is good or bad, what are the interesting features, from where you should buy the product.

Below, we discuss the basic components of an opinion mining engine. Usually, an opinion can be formulated as a tuple O where O is defined as <H, T, S>. Here, H is the *Opinion Holder* who expresses his/her opinion, T is the topic on which the opinion is given, S is the sentiment expressed in the opinion.

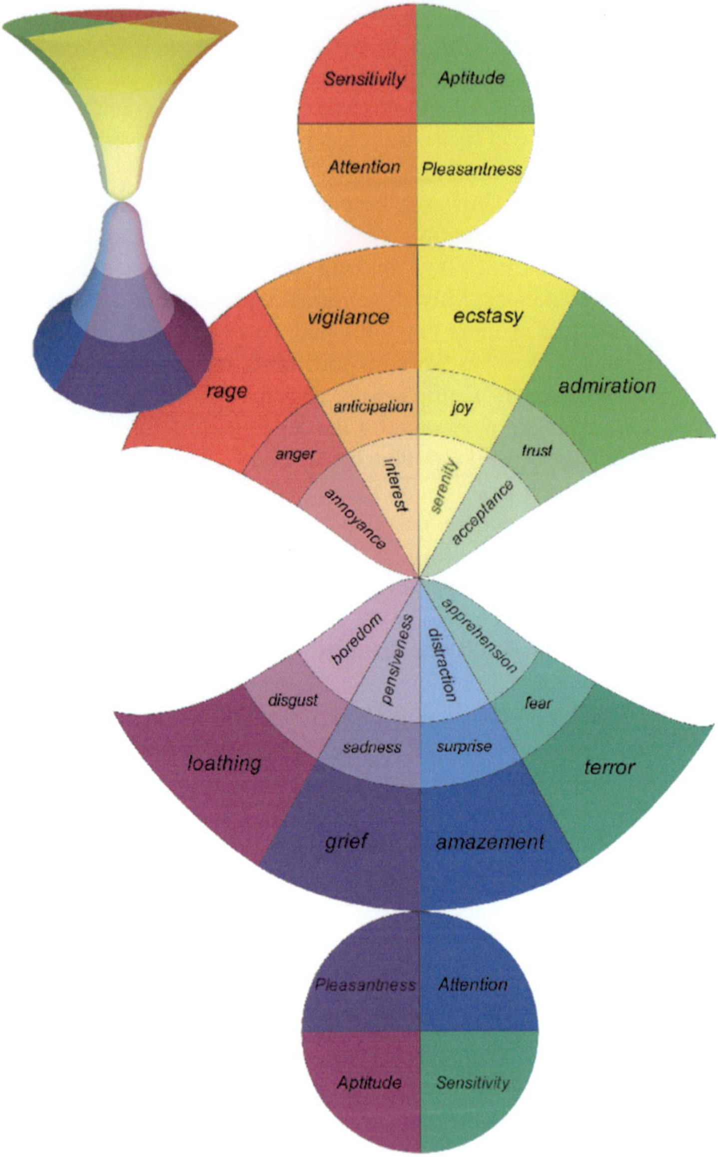

Fig. 2.1 The Hourglass of Emotions

2.2.1 Opinion Holder

Opinion holder is the one who expresses his/her opinion on a product. With the rise of online shopping, the e-commerce websites let people to use *Star rating (1–5)* or *thumbs up – thumbs down*.

Opinion holder detection is a subtask of sentiment analysis and opinion mining. There are cases when more than opinion holder are present in a sentence. Such cases need special treatment in opinion mining. For example, in this sentence *I do not like the movie which you like*, there are two opinion holders and the expressed sentiment is different for each of them.

2.2.2 *Aspects*

Aspects are the product features on which opinion holder expresses her opinion. *The battery life is long*, in this sentence the product feature is *battery life*. The sentiment expressed on this product feature is positive. Aspect extraction from opinions is a subtask of aspect based sentiment analysis. The literature on Aspect level sentiment analysis is discussed in Chap. 3. Aspects can be treated as subtopics if the product itself is considered as a topic. Below in Fig. 2.2 we can see that *iPhone* is the topic of the opinion, *display* and *battery life* are the aspects.

2.2.3 *Subjectivity*

Before analyzing if a sentence is positive or negative, we need to know whether that sentence expresses any sentiment. This is called subjectivity analysis. In sentiment analysis process, we first detect the subjectivity and filters our the objective or neutral ones. The subjective sentences are further classified into positive and negative sentiment categories.

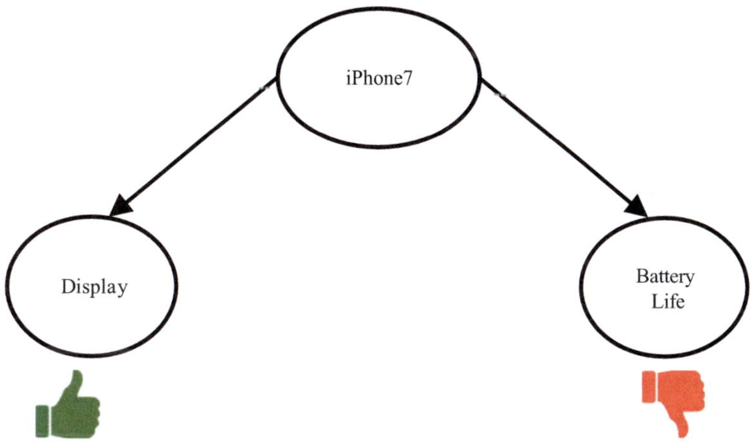

Fig. 2.2 Aspects of the sentence – "iPhone 7 has good display but bad battery life. The sentiment on the aspects are shown using Thumbs up (positive) and Thumbs down (negative)

2.3 Pattern Recognition

In this section, we explain the basic terminologies involved in the topic of Pattern Recognition.

2.3.1 Features

A feature generally corresponds to specific characteristic of a phenomena. One can distinguish a particular type of event or thing based on its features. This also helps to differentiate between two events or products. In the language of computing we often encode such features in numeric to facilitate the computation. The features can be discrete, continuous or can be string.

2.3.2 Pattern

A pattern is structural or quantitative of an object that is usually defined as a form of several features. A pattern helps to distinguish a particular object from other objects.

2.3.3 Class

Before coming to the definition of *class* lets take an example from the sentiment analysis field. If we consider two sentences here –

- The car does not look nice.
- I like this car as it looks awesome.

While the former expresses a negative sentiment on the car, the latter carries a positive sentiment. So, these two sentences correspond to two different classes i.e., positive and negative. Any linguistic as well as computer scientist would say these two sentences carry different patterns. This information helps us to define *class* – a class contains the patterns of similar type but different from the patterns of other classes. Simply, similar features will be found in the samples of a class. The process of grouping the set of patterns into classes is called clustering.

2.4 Feature Selection

In this section we discuss feature selection algorithm Principal Component Analysis (PCA). In machine learning tasks, we need to often process high dimensional data which consume lot of time and space. In order to overcome this problem, numerous feature selection methods were proposed in the literature. PCA [101, 316] is one out of them which we have used in our case. The reason is its simplicity and good performance according to the literature.

2.4.1 *Principal Component Analysis*

Principal Component Analysis (PCA) is a method of dimensionality reduction and feature selection that transforms the data from a d-dimensional space into a new coordinate system of dimension p. Where $p \leq d$ (worst case $p = d$) [171].

The goal of PCA is to preserve as much variance as possible in the new coordinate system. Thus, given the data points on d variables, it is the hope to find a linear subspace where these data points mainly lie in a dimension significantly lower than d. However in practice it is often seen that these data do not precisely lie in some lower dimension subspace.

The new variables that form this new coordinate system are called **Principal Components.** PCA is also used for representation learning. A mode of visualising data, it maps the original points into a new space of lower dimension for more meaningful representation. The direction of the subspace mentioned captures the maximum variance in the data and is called the First Principal Component. Principal components are generally denoted by,

$$\left\langle \mathbf{u_1}, \mathbf{u_2}, \mathbf{u_3}, \ldots, \mathbf{u_d} \right\rangle$$

Since PCs are orthogonal linear transformations of the original variables, therefore there are atmost d PCs. Normally not all of the PCs are used, rather a subset of size p is used. These p-PCs can be chosen based on the percentage of variance that one would like to maintain.

To capture the maximum variability possible, let us choose the first principal component as $\mathbf{u_1}$ and all the data points being stacked into the columns of a matrix $\mathbf{X}$ with dimensions $\mathbf{d} \times \mathbf{n}$.

The projection of these $\mathbf{n}$, $\mathbf{d}-$ dimensional data points on the first principal component $\mathbf{u_1}$ is $\mathbf{u_1^T X}$

To maximise the variance captured by this principal component, we need to maximise:

$$\mathbf{var(u_1^T X) = u_1^T S u_1}$$

where $\mathbf{S}$ is the $\mathbf{d} \times \mathbf{d}$ sample covariance matrix of $\mathbf{X}$.

This optimisation does not have an upper limit because the value can be made arbitrarily large by increasing the magnitude of $\mathbf{u_1}$. Thus we rewrite the optimization including the constraint on the length of $\mathbf{u_1}$ to be unit length:

$$\mathbf{max(u_1^T S u_1)}$$

given,

$$\mathbf{u_1^T u_1 = 1}$$

To solve this optimisation, a Lagrangian multiplier is introduced,

$$\mathbf{L(u_1, \lambda) = u_1^T S u_1 - \lambda(u_1^T u_1 - 1)}$$

Differentiating with respect to $\mathbf{u_1}$, we get,

$$\mathbf{S u_1 = \lambda u_1}$$

Thus,

$$\mathbf{u_1^T S u_1 = \lambda u_1^T u_1 = \lambda}$$

$\mathbf{u_1^T S u_1}$ is maximised if λ is the largest eigenvalue of **S**.

Similarly we can find the rest of the p principal components which captures the remaining maximum variance so that less information is lost in this lower dimensional representation.

PCA, although an age old concept, still prove to be a fundamental tool for dimensionality reduction and feature extraction in numerous machine learning tasks.

2.5 Model Evaluation Techniques

In this section we describe some of the model evaluation techniques to calculate the quality of the regression and classification models.

2.5.1 Evaluating Regression Quality

2.5.1.1 Mean Squared Error Validation Techniques

In Machine Learning, the Mean Squared Error (MSE) or mean squared deviation (MSD) of an estimator measures the average of the squares of the errors or deviations, that is, the difference between the estimator and what is estimated. MSE is a risk function, corresponding to the expected value of the squared error loss or quadratic loss. The difference occurs because of randomness or because the estimator doesn't account for information that could produce a more accurate estimate. If $\hat{Y}$ is a vector of n predictions, and Y is the vector of observed values corresponding to the inputs to the function which generated the predictions, then the MSE of the predictor can be estimated by

$$\mathrm{MSE} = \frac{1}{n}\sum_{i=1}^{n}(\hat{Y}_i - Y_i)^2$$

I.e., the MSE is the mean ($\frac{1}{n}\sum_{i=1}^{n}$) of the square of the errors ($(\hat{Y}_i - Y_i)^2$). This is an easily computable quantity for a particular sample (and hence is sample-dependent).

2.5.2 *Evaluating Classification Techniques*

2.5.2.1 Precision

Precision and recall in information retrieval are defined in terms of retrieved documents i.e., assuming that a search query is made on a search engine, the documents retrieved are used to define precision. Therefore in terms of retrieved documents,precision is the fraction of retrieved documents that are relevant to the query. The formula used for precision is :- $precision = \frac{|(Relavent Documents) \cap (Retrieved Documents)|}{|(Retrieved Documents)|}$

2.5.2.2 Recall

Recall in information retrieval is the fraction of the documents that are relevant to the query that are successfully retrieved. The formula is:-

$$Recall = \frac{|(Relavent Documents) \cap (Retrieved Documents)|}{|(Relavent Documents)|}$$

2.5.2.3 F-Score

A measure that combines precision and recall is the harmonic mean of precision and recall is called the F-Score. The formula:-

$$FScore = \frac{Precision.Recall}{Precision + Recall}$$

2.5.2.4 Accuracy

is a weighted arithmetic mean of Precision and Inverse Precision (weighted by Bias) as well as a weighted arithmetic mean of Recall and Inverse Recall. Accuracy is given by the below formula:-

$$Accuracy = \frac{\sum TruePositive + \sum TrueNegative}{TotalPopulation}$$

2.6 Model Validation Techniques

2.6.1 *Cross Validation*

Cross-validation is a technique to evaluate predictive models by partitioning the original sample into a training set to train the model, and a test set to evaluate it.

In **k-fold cross-validation**, the original sample is randomly partitioned into k equal size subsamples. Of the k subsamples, a single subsample is retained as the validation data for testing the model, and the remaining k-1 subsamples are used as training data. The cross-validation process is then repeated k times (the folds), with each of the k subsamples used exactly once as the validation data. The k results from the folds can then be averaged (or otherwise combined) to produce a single estimation. The advantage of this method is that all observations are used for both training and validation, and each observation is used for validation exactly once.

For classification problems, one typically uses stratified k-fold cross-validation, in which the folds are selected so that each fold contains roughly the same proportions of class labels.

2.6.2 *Bootstrapping*

Bootstrap aggregating, also called bagging, is a machine learning ensemble meta-algorithm designed to improve the stability and accuracy of machine learning algorithms used in statistical classification and regression. It also reduces variance and helps to avoid over fitting. Although it is usually applied to decision tree methods, it can be used with any type of method. Bagging is a special case of the model averaging approach.

Bootstrapping is the practice of estimating properties of an estimator (such as its variance) by measuring those properties when sampling from an approximating distribution. One standard choice for an approximating distribution is the empirical distribution function of the observed data. In the case where a set of observations can be assumed to be from an independent and identically distributed population, this can be implemented by constructing a number of re-samples with replacement, of the observed dataset (and of equal size to the observed dataset).

2.7 Classification Techniques

In this section we first describe what is classification and then we discuss some important classifiers used in this research.

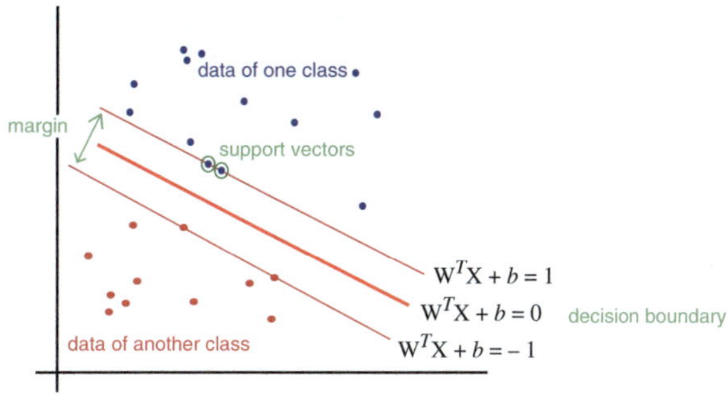

Fig. 2.3 Example of linearly separable data

2.7.1 Support Vector Machine

The simplest form of classification is, the classification of linearly separable data-points with two labels. The diagram below, shows an example of such a dataset.

The goal is to find a hyperplane (also known as *decision boundary*) of the form $\mathbf{w}^\mathsf{T}\mathbf{x} + b = 0$, which separates data-points of two different classes (blue and red).

We can label any point above the decision boundary with 1 and points below with -1. The advantage of this labeling scheme is, the decision function can be concisely written as $f(x) = sign(\mathbf{w}^\mathsf{T}\mathbf{x}+b)$, since $\mathbf{w}^\mathsf{T}\mathbf{x}+b > 0$ for all points above the hyperplane and $\mathbf{w}^\mathsf{T}\mathbf{x} + b < 0$ for all points below.

Now, you might notice that, there is a gap between the decision boundary and the nearest point(s) (also known as *support vectors*) of either of the classes. We need to choose the hyperplane in such a way that, the gap is maximized. Such a hyperplane is called *Maximal Marginal Hyperplane* (MMH). The idea is to keep the classes as far as possible from the decision boundary to minimize classification error (Fig. 2.3).

Let, $\mathbf{w}^\mathsf{T}\mathbf{x} + b = 1$ be a hyperplane, such that all points on and above it belong to class +1. Similarly, $\mathbf{w}^\mathsf{T}\mathbf{x} + b = -1$ be a hyperplane, such that all points on and below it belong to class -1. So, we can check if a data-point $\mathbf{x_i}$, labeled $y_i(= +1$ or $-1)$, has been correctly classified by verifying if $y_i(\mathbf{w}^\mathsf{T}\mathbf{x_i} + b) \geq 1$.

Hyperplanes $\mathbf{w}^\mathsf{T}\mathbf{x} + b = 1$ and $\mathbf{w}^\mathsf{T}\mathbf{x} + b = -1$ are parallel, since they have the same normal vector given by $\mathbf{w}$. So, we need to maximize the distance between the given hyperplanes.

The distance between the hyperplanes can be given by $\dfrac{|(b-1)-(b+1)|}{||\mathbf{w}||} = \dfrac{2}{||\mathbf{w}||}$. Hence, the objective is to maximize $\dfrac{2}{||\mathbf{w}||}$. It is equivalent to minimizing $\dfrac{||\mathbf{w}||}{2} = \dfrac{\sqrt{\mathbf{w}^\mathsf{T}\mathbf{w}}}{2}$. Again, it is equivalent to minimizing $\dfrac{\mathbf{w}^\mathsf{T}\mathbf{w}}{2}$.

Finally, the optimization problem can be summarized as:

$$min_{\mathbf{w},b} \frac{\mathbf{w}^{\mathsf{T}}\mathbf{w}}{2}$$

subject to $y_i(\mathbf{w}^{\mathsf{T}}\mathbf{x_i} + b) \geq 1$, for all $i = 1, 2 \ldots, m$.

2.7.1.1 Soft-Margin Extension

Now, consider that the data-points are not completely linearly separable. In such case, we need to allow some data-points to lie on the other side of the decision boundary.

We introduce *slack variables* $\zeta_i \geq 0$ for all x_i and rewrite the problem as follows:

$$min_{\mathbf{w},b,\zeta} \frac{\mathbf{w}^{\mathsf{T}}\mathbf{w}}{2} + C\sum_{i=1}^{m} \zeta_i$$

subject to $y_i(\mathbf{w}^{\mathsf{T}}\mathbf{x_i} + b) \geq 1 - \zeta_i,\ \zeta_i \geq 0$ for all $i = 1, 2 \ldots, m$.

Slack variables allow the quantity $y_i(\mathbf{w}^{\mathsf{T}}\mathbf{x_i} + b)$ to be less than one; meaning $\mathbf{x_i}$ lies on the other side of the decision boundary. However, we penalize such occurrences by adding slack variables to the objective function.

2.7.1.2 Non-linear Decision Boundary

The data-points $\mathbf{x_i}$ may not be linearly separable in the original space. However, if we map the data-points to a higher dimensional space, it might be linearly separable in that new space.

We use a mapping function ϕ to map the data-points. Hence, we rewrite the optimization problem as follows:

$$min_{\mathbf{w},b,\zeta} \frac{\mathbf{w}^{\mathsf{T}}\mathbf{w}}{2} + C\sum_{i=1}^{m} \zeta_i$$

subject to $y_i(\mathbf{w}^{\mathsf{T}}\phi(\mathbf{x_i}) + b) \geq 1 - \zeta_i,\ \zeta_i \geq 0$ for all $i = 1, 2 \ldots, m$.

Example of a mapping function:

$$\phi : \mathbb{R}^2 \mapsto \mathbb{R}^3$$

$$(x_1, x_2) \rightarrow (z_1, z_2, z_3) := (x_1^2, \sqrt{2}x_1x_2, x_2^2)$$

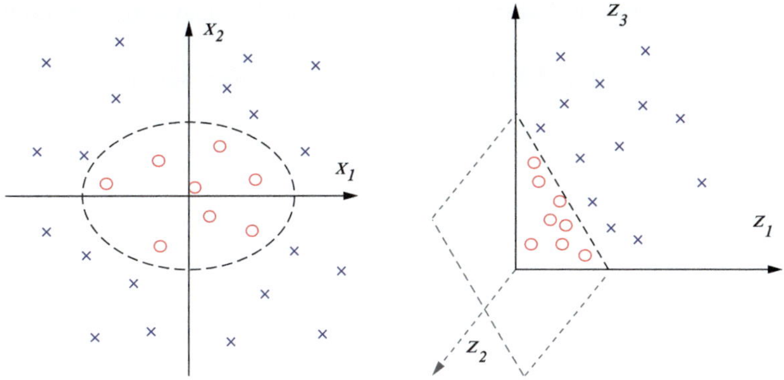

Fig. 2.4 Polynomial mapping

2.7.1.3 Formulation as a Lagrangian Optimization

We can use *Lagrangian Multiplier* to incorporate the constraints.

Now, $\zeta_i \geq 1 - y_i(\mathbf{w}^\mathsf{T}\phi(\mathbf{x_i}) + b)$. So, in order for ζ_i to be near zero, $1 - y_i(\mathbf{w}^\mathsf{T}\phi(\mathbf{x_i}) + b)$ has to be as close to zero as possible (Fig. 2.4).

We can capture this condition by adding the following to the objective function: $max_{\alpha_i \geq 0}\alpha_i[1 - y_i(\mathbf{w}^\mathsf{T}\phi(\mathbf{x_i}) + b)]$. If $y_i\mathbf{w}^\mathsf{T}\phi(\mathbf{x_i}) + b \geq 1$, then α_i becomes zero; since $1 - y_i(\mathbf{w}^\mathsf{T}\phi(\mathbf{x_i}) + b)$ becomes negative. Otherwise, $\alpha_i \to \infty$. In this way, we are penalizing misclassified points, since $max_{\alpha_i \geq 0}\alpha_i[1 - y_i(\mathbf{w}^\mathsf{T}\phi(\mathbf{x_i}) + b)]$ becomes positive and huge.

So, we rewrite the objective function as:

$$min_{\mathbf{w},b}[\frac{\mathbf{w}^\mathsf{T}\mathbf{w}}{2} + \sum_{i=1}^{m} max_{\alpha_i \geq 0}\alpha_i[1 - y_i(\mathbf{w}^\mathsf{T}\phi(\mathbf{x_i}) + b)]].$$

To allow soft-margin, we limit the value of α_i to lie within $[0, C]$.

Now, we can rewrite the formulation as below:

$$\begin{aligned}
&min_{\mathbf{w},b}[\frac{\mathbf{w}^\mathsf{T}\mathbf{w}}{2} + \sum_{i=1}^{m} max_{\alpha_i \geq 0}\alpha_i[1 - y_i(\mathbf{w}^\mathsf{T}\phi(\mathbf{x_i}) + b)]]\\
&= min_{\mathbf{w},b}[max_{\alpha \geq 0}[\frac{\mathbf{w}^\mathsf{T}\mathbf{w}}{2} + \sum_{i=1}^{m} \alpha_i[1 - y_i(\mathbf{w}^\mathsf{T}\phi(\mathbf{x_i}) + b)]]]\\
&= min_{\mathbf{w},b}[max_{\alpha \geq 0}J(\mathbf{w}, b, \alpha)], \text{ where } J(\mathbf{w}, b, \alpha)\\
&= \frac{\mathbf{w}^\mathsf{T}\mathbf{w}}{2} + \sum_{i=1}^{m} \alpha_i[1 - y_i(\mathbf{w}^\mathsf{T}\phi(\mathbf{x_i}) + b)]\\
&= max_{\alpha \geq 0}[min_{\mathbf{w},b}J(\mathbf{w}, b, \alpha)]
\end{aligned}$$

So, now we minimize J with respect to $\mathbf{w}$, b and then maximize the result with respect to α.

Now, we set $\frac{\partial J}{\partial \mathbf{w}} = 0$, and find $\mathbf{w} = \sum_i \alpha_i y_i \phi(\mathbf{x_i})$. Again, setting $\frac{\partial J}{\partial b} = 0$, we get the relation $\sum_i \alpha_i y_i = 0$.

After, substitution of the above mentioned values and relation, we get:

$$min_{\mathbf{w},b} J(\mathbf{w}, b, \alpha) = \sum_i \alpha_i - \frac{1}{2} \sum_{i,j} \alpha_i \alpha_j y_i y_j \phi(\mathbf{x_i})^\mathsf{T} \phi(\mathbf{x_j}).$$

Thus, finally, the dual problem is:

$$max_{\alpha \geq 0} [\sum_i \alpha_i - \frac{1}{2} \sum_{i,j} \alpha_i \alpha_j y_i y_j \phi(\mathbf{x_i})^\mathsf{T} \phi(\mathbf{x_j})]$$

Subject to $\sum_i \alpha_i y_i = 0, \; 0 \leq \alpha_i \leq C$

2.7.1.4 Kernel Trick

Since, we are mapping the data-points to a very high-dimensional space, calculating the term $\phi(\mathbf{x_i})^\mathsf{T} \phi(\mathbf{x_j})$ may become intractable. However, we can avoid the dot-product by introducing special *kernel functions*, which evaluates the dot-product by operating on the original lower-dimension.

For example, let $\phi : \mathbb{R}^3 \mapsto \mathbb{R}^{10}$, where:

$$\phi(\mathbf{x}) = (1, \sqrt{2}x_1, \sqrt{2}x_2, \sqrt{2}x_3, x_1^2, x_2^2, x_3^2, \sqrt{2}x_1 x_2, \sqrt{2}x_2 x_3, \sqrt{2}x_1 x_3)$$

Now, we can use kernel function $K(\mathbf{x_i}, \mathbf{x_j}) = \phi(\mathbf{x_i})^\mathsf{T} \phi(\mathbf{x_j}) = (1 + \mathbf{x_i}^\mathsf{T} \mathbf{x_j})^2$.

So, we can rewrite the dual problem as:

$$max_{\alpha \geq 0} [\sum_i \alpha_i - \frac{1}{2} \sum_{i,j} \alpha_i \alpha_j y_i y_j K(\mathbf{x_i}, \mathbf{x_j})]$$

Subject to $\sum_i \alpha_i y_i = 0, \; 0 \leq \alpha_i \leq C$

Decision Function

We learn the parameters α and b. Then, we classify a test instance $\mathbf{x}$, by calculating $f(\mathbf{x}) = sign(\mathbf{w}^\mathsf{T} \phi(\mathbf{x}) + b)$. We substitute $\mathbf{w} = \sum_i \alpha_i y_i \phi(\mathbf{x_i})$ and get $f(x) = sign(\sum_i \alpha_i y_i \phi(\mathbf{x_i})^\mathsf{T} \phi(\mathbf{x}) + b) = sign(\sum_i \alpha_i y_i K(\mathbf{x_i}, \mathbf{x}) + b)$.

2.7.2 Extreme Learning Machine

The ELM approach [155] was introduced to overcome some issues in back-propagation network [321] training, specifically, potentially slow convergence rates, the critical tuning of optimization parameters [378], and the presence of local minima that call for multi-start and re-training strategies. The ELM learning problem settings require a training set, X, of N labeled pairs, where $(\mathbf{x}_i, y_i)$, where $\mathbf{x}_i \in \mathcal{R}^m$ is the i-th input vector and $y_i \in \mathcal{R}$ is the associate expected 'target' value; using a scalar output implies that the network has one output unit, without loss of generality.

The input layer has m neurons and connects to the 'hidden' layer (having N_h neurons) through a set of weights $\{\hat{\mathbf{w}}_j \in \mathcal{R}^m; j = 1, \ldots, N_h\}$. The j-th hidden neuron embeds a bias term, $\hat{b}_j$,and a nonlinear 'activation' function, $\varphi(\cdot)$; thus the neuron's response to an input stimulus, $\mathbf{x}$, is:

$$a_j(\mathbf{x}) = \varphi(\hat{\mathbf{w}}_j \cdot \mathbf{x} + \hat{b}_j) \tag{2.1}$$

Note that (2.1) can be further generalized to a wider class of functions [154] but for the subsequent analysis this aspect is not relevant. A vector of weighted links, $\bar{\mathbf{w}}_j \in \mathcal{R}^{N_h}$, connects hidden neurons to the output neuron without any bias [151]. The overall output function, $f(\mathbf{x})$, of the network is:

$$f(\mathbf{x}) = \sum_{j=1}^{N_h} \bar{\mathbf{w}}_j a_j(\mathbf{x}) \tag{2.2}$$

It is convenient to define an 'activation matrix', $\mathbf{H}$, such that the entry $\{h_{ij} \in \mathbf{H}; i = 1, \ldots, N; j = 1, \ldots, N_h\}$ is the activation value of the j-th hidden neuron for the i-th input pattern. The $\mathbf{H}$ matrix is:

$$\mathbf{H} \equiv \begin{bmatrix} \varphi(\hat{\mathbf{w}}_1 \cdot \mathbf{x}_1 + \hat{b}_1) & \cdots & \varphi(\hat{\mathbf{w}}_{N_h} \cdot \mathbf{x}_1 + \hat{b}_{N_h}) \\ \vdots & \ddots & \vdots \\ \varphi(\hat{\mathbf{w}}_1 \cdot \mathbf{x}_N + \hat{b}_1) & \cdots & \varphi(\hat{\mathbf{w}}_{N_h} \cdot \mathbf{x}_N + \hat{b}_{N_h}) \end{bmatrix} \tag{2.3}$$

In the ELM model, the quantities $\{\hat{\mathbf{w}}_j, \hat{b}_j\}$ in (2.1) are set randomly and are not subject to any adjustment, and the quantities $\{\bar{\mathbf{w}}_j, \bar{b}\}$ in (2.2) are the only degrees of freedom. The training problem reduces to the minimization of the convex cost:

$$\min_{\{\bar{\mathbf{w}}, \bar{b}\}} \left\| \mathbf{H}\bar{\mathbf{w}} - \mathbf{y} \right\|^2 \tag{2.4}$$

A matrix pseudo-inversion yields the unique L_2 solution, as proven in [155]:

$$\bar{\mathbf{w}} = \mathbf{H}^+ \mathbf{y} \tag{2.5}$$

The simple, efficient procedure to train an ELM therefore involves the following steps:

1. Randomly set the input weights $\hat{\mathbf{w}}_i$ and bias $\hat{b}_i$ for each hidden neuron;
2. Compute the activation matrix, $\mathbf{H}$, as per (2.3);
3. Compute the output weights by solving a pseudo-inverse problem as per (2.5).

Despite the apparent simplicity of the ELM approach, the crucial result is that even random weights in the hidden layer endow a network with a notable representation ability [155]. Moreover, the theory derived in [156] proves that regularization strategies can further improve its generalization performance. As a result, the cost function (2.4) is augmented by an L_2 regularization factor as follows:

$$\min_{\bar{\mathbf{w}}} \{\left\|\mathbf{H}\bar{\mathbf{w}} - \mathbf{y}\right\|^2 + \lambda\left\|\bar{\mathbf{w}}\right\|^2\} \tag{2.6}$$

2.7.3 Deep Neural Networks

A deep neural network (DNN) can be viewed as a composite of simple, unsupervised models such as Restricted Boltzmann machines (RBMs) where each hidden layer serves as the visible layer for the next RBM. RBM is a bipartite graph comprising two layers of neurons: a visible and a hidden layer; it is restricted such that the connections among neurons in the same layer are not allowed.

To train such a multi-layer system, we must compute the gradient of the total energy function E with respect to weights in all the layers. To learn such weights and maximize the global energy function, the approximate maximum likelihood contrastive divergence approach can be used. This method employs each training sample to initialize the visible layer. Next, it uses the Gibbs sampling algorithm to update the hidden layer and then reconstruct the visible layer consecutively, until convergence. As an example, here we use a logistic regression model to learn the binary hidden neurons and each visible unit is assumed to be a sample from a normal distribution. The continuous state $\hat{h}_j$ of the hidden neuron j, with bias b_j, is a weighted sum over all continuous visible nodes $\boldsymbol{v}$ and is given by:

$$\hat{h}_j = b_j + \sum_i v_i w_{ij}, \tag{2.7}$$

where w_{ij} is the connection weight to hidden neuron j from visible node v_i. The binary state h_j of the hidden neuron can be defined by a sigmoid activation function:

$$h_j = \frac{1}{1 + e^{-\hat{h}_j}}, \tag{2.8}$$

Similarly, in the next iteration, the continuous state of each visible node v_i is reconstructed. Here, we determine the state of visible node i, with bias c_i, as a random sample from the normal distribution where the mean is a weighted sum over all binary hidden neurons and is given by:

$$v_i = c_i + \sum_j h_i w_{ij}, \tag{2.9}$$

where w_{ij} is the connection weight to hidden neuron j from visible node i. This continuous state is a random sample from $\mathcal{N}(v_i, \sigma)$, where σ is the variance of all visible nodes.

Unlike hidden neurons, visible nodes can take continuous values in a Gaussian RBM. Lastly, the weights are updated as the difference between the original and reconstructed visible layer labelled as the vector $\boldsymbol{v}_{recon}$ using:

$$\Delta w_{ij} = \alpha(< v_i h_j >_{data} - < v_i h_j >_{recon}), \tag{2.10}$$

where α is the learning rate and $< v_i h_j >$ is the expected frequency with which visible unit i and hidden unit j are active together when the visible vectors are sampled from the training set and the hidden units are determined by (2.7). Finally, the energy of a DNN can be determined in the final layer using:

$$\mathrm{E} = -\sum_{i,j} v_i h_j w_{ij}, \tag{2.11}$$

To extend the DNN to convolutional deep neural network, we simply partition the hidden layer into Z groups. Each of the Z groups is associated with a $n_x \times n_y$ filter where n_x is the height of the kernel and n_y is the width of the kernel. Let us assume that the input image has dimension $\mathrm{L_x} \times \mathrm{L_y}$. Then the convolution will result in a hidden layer of Z groups each of dimension $(\mathrm{L_x} - n_x + 1) \times (\mathrm{L_y} - n_y + 1)$. These learned kernel weights are shared among all hidden units in a particular group. The energy function of layer l is now a sum over the energy of individual blocks given by:

$$\mathrm{E}^l = -\sum_{z=1}^{\mathrm{Z}} \sum_{i,j}^{(\mathrm{L_x}-n_x+1),(\mathrm{L_y}-n_y+1)} \sum_{r,s}^{n_x,n_y} v_{i+r-1,j+s-1} h_{ij}^z w_{rs}^l. \tag{2.12}$$

It can be seen that in deep convolutional learning of images, pixels combine to form edge detectors in the first layer, and edge detectors combine to form bigger image parts in the second layer and so on.

2.8 Feature-Based Text Representation

Word Embedding is an important technique in NLP to represent words (atomic unit of text) as a real valued vector of relatively low dimension as compared to the vocabulary size. Originally presented in 2003 by Bengio et al., this mapping explicitly uses neural networks, deep learning, dimensionality reduction and probabilistic models to extract both syntactic and semantic contexts in the vector space. Word Embeddings have proven to be efficient in capturing context similarity, analogies and due to its smaller dimensionality, are fast and efficient in computing core NLP tasks.

Mainly associated with text representation, word embeddings have proven to be very efficient in capturing both syntactic and semantic concepts from a given corpus which can be further used in numerous other NLP tasks like sentimental analysis, named entity recognition to name a few.

Words are considered as the most basic unit in semantic tasks. Thus, their representation in being used as input to NLP models is of prime importance. Having said that, there are also tasks which delve into finer grains of defining characters as the atomic unit. Tasks like Parts of Speech Tagging or Lemmatization often look into the characters as its prime input. Despite this distinction, this article focuses on the numerous tasks that uses words to be their input and deals with the challenges in representing these words given a corpus with a huge vocabulary (Fig. 2.5).

It is well known that computers are very good with numbers and in contrast, raw text calls for string operations which can turn out to be very costly. Thus, representing words as vectors of numbers is preferred. There are different ways in which a word can be represented. The following sections introduce these methods followed by brief discussion on probabilistic models used in creating these word embeddings.

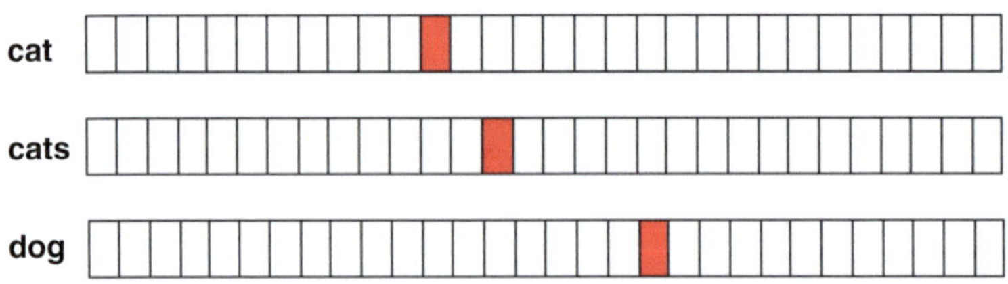

Fig. 2.5 One hot encoding

2.8.1 Types of Feature-Based Text Representation

2.8.1.1 One-Hot Vectors

It is the most trivial way of representing a word into a binary vector v of length equal to the size of the vocabulary, i.e., $len(v) = \|V\|$ where V is the vocabulary of unique words in the corpus. In this vector, $v_k = 1$, where k is the index of the word in the vocabulary and $\forall k^{'} \neq k, v_{k'} = 0$.

Main problem with one-hot vectors lies in its sparseness which fails to capture any semantic similarity between words. Also their dimension being the size of the vocabulary is too large for larger vocabularies.

This problem can be tackled by using dimensionality reduction methods like **SVD** for linear reduction and **LLE**, Locally Linear Embedding for non- linear reduction into a subspace of smaller dimension.

Computational costs in SVDs is high as it scales quadratically i.e. $O(nd^2)$ for a matrix of dimension $d \times n$. Thus better method would be to learn and compute low-dimensional vectors directly.

2.8.1.2 Distributional Vectors

The essence of distributional vectors is derived from the Distributional Hypothesis which states that: *Words that are used and occur in the same context tend to support similar meanings.*

Thus, these vectors try to capture the characteristics of the neighbours of a word. Each word is again represented by a $\|V\|$- dimensional vector but this time, each entry in this vector is the frequency count of the corresponding indexed word being a neighbour to the target word within a certain distance k, known as the window size (Fig. 2.6).

The main advantage of distributional vectors is that they capture similarity between words: similar words → similar neighbors → similar vectors. Measuring similarity between vectors is possible, using measures such as cosine similarity, for example.

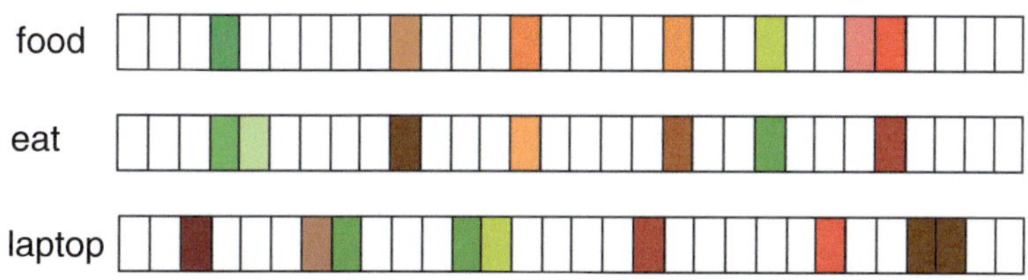

Fig. 2.6 Distributional Vector

2.8.1.3 Word Embeddings

This method is able to directly calculate the contextual information in a low-dimensional vector. Each word is now represented by a small dimension D, generally in the range of 50–1000.

The basic difference lies in the fact that co-occurrences are no longer counted in the window but rather predicted. The vectors for each word is randomly initialised and then each word is iterated in the whole corpus. At every step, the vectors of this target word and all its context words (words that belong to the current window) are updated to bring them closer in the D-dimensional space. Thus, according to the distributional hypothesis, semantically similar words should be pulled (updated) closer in this D-dimensional space.

The advantages of word embeddings lie in its relatively low dimensions while keeping the information of word similarity intact. Due to its compactness, computations on these vectors are efficient and fast.

2.8.2 Word2Vec

Word embeddings were revolutionised when Mikolov et al. (2013) introduced Word2Vec [232, 233]. Interestingly the model used here was a shallow neural network which was trained to learn these vectors from the corpora of complete Wikipedia. Word2vec has two models, namely **CBOW** and **SkipGram**. CBOW model, given some words in a context belonging to a window of size k, computes the most probable target word using the conditional probability of the target word given the context words. On the other hand, the SkipGram model does the exact opposite of the CBOW model, by predicting the surrounding context words given the central target word. The context words here are assumed to be located symmetrically to the target words within a distance equal to the window size in both directions (Fig. 2.7).

To understand the CBOW model, we start with the simplest version of this model where only one word is considered in the context. This means that the model has to predict a target word given a context word. The settings here replicate a bigram language model.

Let us assume a single hidden layer neural network with **V** neurons in the input and output layer and **H** number of neurons in the hidden layer. The layers are fully connected and input to the network is the context word represented as a one-hot vector. The layers are connected by weight matrix $\mathbf{W_{V\times H}}$ and $\mathbf{W'_{H\times V}}$ respectively. Here, each word is finally represented as two learned vectors $\mathbf{V_c}$ and $\mathbf{V_w}$, representing their roles as context and target words respectively. Given an input one-hot vector **x**, the hidden layer can be calculated as follows

$$\mathbf{h} = \mathbf{W^T x} = \mathbf{W_{(k,.)}}$$

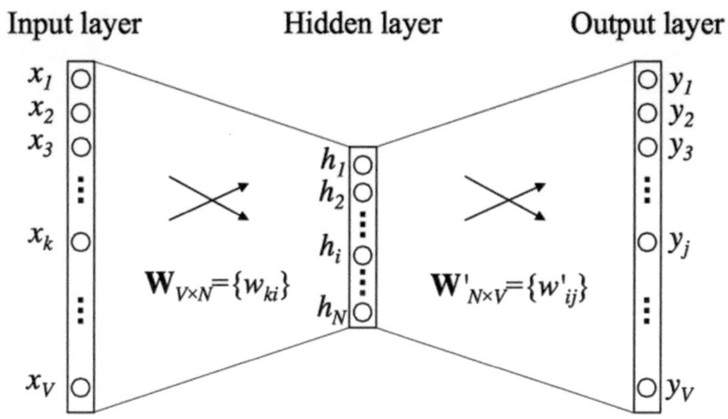

Fig. 2.7 CBOW architecture

Thus this **kth** row of **W** represents the output vector $\mathbf{V_c}$ for the input vector **x**. Also for each output unit let,

$$\mathbf{y_i} = softmax(\mathbf{u_i})$$

$$where, \mathbf{u} = \mathbf{W}^{'T}\mathbf{h}$$

Thus,

$$\mathbf{u_i} = \mathbf{W}^{'}_{(.,\mathbf{i})}\mathbf{h}$$

The **kth** column of matrix $\mathbf{W}^{'}$ therefore represents the output vector $\mathbf{V_w}$.

Overall,

$$\mathbf{P}(\frac{\mathbf{w_i}}{\mathbf{c}}) = \mathbf{y_l} = \frac{e^{u_i}}{\sum_{i=1}^{V} e^{u_i}}$$

$$where, \mathbf{u_i} = \mathbf{V}_{\mathbf{w_i}}^{\mathbf{T}}\mathbf{V_c}$$

The parameters $\boldsymbol{\theta} = \{\mathbf{V_w}, \mathbf{V_c}\}$ are learnt by defining the objective function as the log-likelihood,

$$\mathbf{l}(\theta) = \sum_{\mathbf{w}\in\mathbf{Vocabulary}} \mathbf{log}(\mathbf{P}(\frac{\mathbf{w}}{\mathbf{c}}))$$

$$= \sum_{\mathbf{w}\in\mathbf{Vocabulary}} \mathbf{log}(\frac{\mathbf{e}^{\mathbf{V}_\mathbf{w}^\mathbf{T}\mathbf{V_c}}}{\sum_{\mathbf{i}=\mathbf{1}}^{\mathbf{V}} \mathbf{e}^{\mathbf{V}_{\mathbf{w_i}}^{\mathbf{T}}\mathbf{V_c}}})$$

Finding the gradient,

$$\frac{\partial \mathbf{l}(\theta)}{\partial \mathbf{V_w}} = \mathbf{V_c} - \frac{\mathbf{e}^{\mathbf{V_w^T V_c}} \mathbf{V_c}}{\sum_{\mathbf{w'} \in \mathbf{Vocab}} \mathbf{e}\mathbf{w}'^{\mathbf{T}} \mathbf{V_c}}$$

$$\equiv \frac{\partial \mathbf{l}(\theta)}{\partial \mathbf{V_w}} = \mathbf{V_c}(\mathbf{1} - \mathbf{P}(\frac{\mathbf{w}}{\mathbf{c}}))$$

After this, updation of weights is done using Gradient Descent:

$$\mathbf{V_w^{new}} = \mathbf{V_w^{old}} - \eta \frac{\partial \mathbf{l}(\theta)}{\partial \mathbf{V_w}}$$

Similarly doing for $\mathbf{V_c}$.

In the General Model, all the one-hot vectors of context words are taken as input simultaneously, i.e.,

$$\mathbf{h} = \mathbf{W^T}(\mathbf{x_1} + \mathbf{x_2} + \ldots + \mathbf{x_c}).$$

2.8.2.1 Negative Sampling

It is observable here that in the formulae mentioned above, the normalisation factor iterates over the whole vocabulary for each update. This is therefore intractable when the vocabulary is very large. To overcome this, a different approach is taken where the task considers pairs of target and context words and breaks down to classify where this pair belongs to the corpus or not (wrt. window size restrictions).

2.8.2.2 Classification Problem

Let, $\mathbf{D} \rightarrow$ Set of all (w,c) pair of words in the corpus. and, $\mathbf{D}' \rightarrow$ Set of all (w,c) pair of words **not** in the corpus.

$\mathbf{P}(\mathbf{z} == \mathbf{1}|(\mathbf{w}, \mathbf{c}))$ is the probability that a pair is in $\mathbf{D}$

Therefore,

$$\mathbf{P}(\mathbf{z} == \mathbf{1}|(\mathbf{w}, \mathbf{c})) = \frac{\mathbf{1}}{\mathbf{1} + \mathbf{e}^{-\mathbf{V_c^T V_w}}}$$

$$\mathbf{P}(\mathbf{z} == \mathbf{0}|(\mathbf{w}, \mathbf{c})) = \frac{\mathbf{e}^{-\mathbf{V_c^T V_w}}}{\mathbf{1} + \mathbf{e}^{-\mathbf{V_c^T V_w}}}$$

now,

$$\mathbf{P(z|(w,c))} = \Big(\frac{\mathbf{1}}{\mathbf{1+e^{-V_c^TV_w}}}\Big)^{\mathbf{z}}\Big(\frac{\mathbf{e^{-V_c^TV_w}}}{\mathbf{1+e^{-V_c^TV_w}}}\Big)^{\mathbf{1-z}}$$

Hence the Likelihood can be defined as,

$$\mathbf{L}(\theta) = \prod_{(w,c)\in D\cup D'} P(z|(w,c))$$

Defining the log-likelihood as the objective function to minimise:

$$\mathbf{l}(\theta) = \sum_{\mathbf{(w,c)\in D\cup D'}} \mathbf{zlog}(\frac{\mathbf{1}}{\mathbf{1+e^{-V_c^TV_w}}})$$

$$+\mathbf{(1-z)log}(\frac{\mathbf{1}}{\mathbf{1+e^{V_c^TV_w}}})$$

$$\mathbf{l}(\theta) = \sum_{\mathbf{(w,c)\in D}} \mathbf{log}(\frac{\mathbf{1}}{\mathbf{1+e^{-V_c^TV_w}}}) + \sum_{\mathbf{(w,c)\in D'}} \mathbf{log}(\frac{\mathbf{1}}{\mathbf{1+e^{V_c^TV_w}}})$$

From this we can find the respective gradients and perform the required updates using Gradient Descent. The computational complexity in the above model is no more intractable.

Although **Word2Vec** and similar counterparts like **GloVe** [272] are very efficient in capturing semantic details of words in their vector space representations, they are not very efficient in the task of sentimental analysis. This is mainly due to the fact that these embeddings sometimes cluster semantically similar words which have opposing sentiment polarity. Thus the downstream model used for the sentiment analysis task would not be able to identify this contrasting polarities leading to poor performance. Many solutions to this problem have been proposed based on the modification of the original C&W model by Collobert et al. [71]

This is an open field which high scope of improvement in the said task.

2.9 Conclusion

In this chapter, we have discussed the basic theoretical details of sentiment analysis (e.g., aspects, opinions), machine learning classifiers (e.g., SVM, ELM, Deep Learning) and some general text analysis methods (e.g., Word2Vec). In the next chapters, we proposed text-based and multimodal frameworks which we have built upon the theoretical methods discussed in this chapter.

Chapter 3
Literature Survey and Datasets

Abstract In this chapter we present the literature on unimodal and multimodal approaches to sentiment analysis and emotion recognition. As discussed in the Sect. 2.1, both of these topics can be brought together under the umbrella of affective computing.

Keywords Fusion · Multimodal · Text sentiment analysis · Affective computing · Early fusion

3.1 Introduction

In this book, we focus on these two kinds of affect recognition by specifically identifying datasets (Tables 3.1, and 3.2) and works (Tables 3.5, and 3.6) in both fields.

3.2 Available Datasets

Since the main purpose of this chapter is to inform readers about recent advances in multimodal affect recognition, in this section we describe widely-used datasets for multimodal emotion recognition and sentiment analysis. However, we do not cover the unimodal datasets for example facial expression recognition from image datasets (e.g., CK++) as it is outwith the scope of the book.

In the literature, we found two main methodologies for dataset collection: natural videos and video recordings of subjects acting based on pre-decided scripts. To curate the latter, subjects were provided affect-related scripts and asked to act. It is observed that such datasets can suffer from inaccurate actions by subjects, leading to corrupted samples or inaccurate information for the training dataset. According to D'Mello et al. [92], even though in the literature most of the times a multimodal framework achieved performance improvement over unimodal systems, that improvement was much lower when it was trained on natural data (4.59% improvement) versus acted data (12.7% improvement).

S. Poria et al., *Multimodal Sentiment Analysis*, Socio-Affective Computing 8,
https://doi.org/10.1007/978-3-319-95020-4_3

Table 3.1 Multimodal emotion analysis datasets

Dataset	References	Modality	Speakers	Features	Sentiment	Annotators	Availability
HUMAINE	Cowie et al. [97]	A + V	50	Emotion words, authenticity, core affect dimensions, context labels	NA	6 (only 16 clips labeled)	Publicly available http://emotion-research.net/download/pilot-db
Belfast database	Cowie et al. [96]	A + V	125 (31 M, 94 F)	Wide range of emotions	NA	7	On registration http://belfast-naturalistic-db.sspnet.eu
SEMAINE	McKeown et al. [226]	A + V	150	Angry, happy, fear, disgust, disgust, sadness, contempt and amusement	NA	6	On registration http://semaine-db.eu
IEMOCAP	Busso et al. [33]	A + V	10 (5 M, 5 F)	Happiness, anger, sadness, frustration and neutral state	NA	3 for each emotion category	On request http://sail.usc.edu/iemocap
eNTERFACE	Martin et al. [223]	A + V	42 (34 M, 8 F)	Happiness, anger, sadness, surprise, disgust and fear	NA	2	Publicly available http://enterface.net

Legenda: *A* Audio, *V* Video

Table 3.2 Multimodal sentiment analysis datasets

Dataset	References	Modality	Speakers	Features	Sentiment	Annotators	Availability
ICT-MMMO	Wollmer et al. [398]	A + T + V	370	1000 linguistic +1941 acoustic +20 visual	Strongly negative, weakly negative, neutral, weakly positive and strongly positive	3	By sending mail to Giota Stratou (stratou@ict.usc.edu)
MOUD	Rosas et al. [273]	A + T + V	80 (65 F, 15 M)	28 acoustic +40 visual	Positive, negative, and neutral	2	Publicly available http://web.eecs.umich.edu/mihalcea/downloads.html
YouTube dataset	Morency et al. [239]	A + T + V	47 (20 F, 27 M)	1000 linguistic +1941 acoustic +20 visual	Polarized words, smile, look away, Pauses and Pitch	3	By sending mail to Giota Stratou (stratou@ict.usc.edu)

Legenda: *A* Audio, *T* Text, *V* Video

There are several other drawbacks associated with this method, e.g., the time taken to create the dataset and biased labeling. Due to these problems, a model trained on these datasets may suffer from poor generalization capability. To overcome such problems, Morency et al. [239] proposed a method of dataset collection in which the product review videos were crawled from popular social websites and later labeled with emotion and sentiment labels.

A common feature amongst both approaches is that they are labeled at utterance level, i.e., for each utterance there is an associated emotion or sentiment label. Utterance-level labeling scheme is particularly important to track the emotion and sentiment dynamics of the subject's mindset in a video.

3.2.1 Datasets for Multimodal Sentiment Analysis

Available datasets for multimodal sentiment analysis have mostly been collected from product reviews available on different online video sharing platforms, e.g., YouTube. The publicly available multimodal emotion recognition and sentiment analysis datasets are summarized in Tables 3.1 and 3.2, respectively.

3.2.1.1 YouTube Dataset

This dataset was developed in 2011 by Morency et al. [239]. The idea behind its development is to capture the data present in the increasing number of videos posted online every day. The authors take pride in developing the first publicly available dataset for tri-modal sentiment analysis, by combining visual, audio and textual modalities. The dataset was created by collecting videos from YouTube that is diverse, multimodal and has ambient noises. The keywords used for the collection of videos are opinion, review, best perfume, tooth paste, business, war, job, I hate and I like. Finally, a dataset of 47 videos were created, out of which 20 were from female speakers and the rest male, with their ages ranging from 14 to 60 years. All speakers expressed their views in English and belonged to different cultures. The videos were set to .mp4 format with a size of 360×480. The 47 videos in the dataset were further annotated with one of three sentiment labels: positive, negative or neutral. This annotation task led to 13 positively, 12 negatively and 22 neutrally labeled videos. For qualitative and statistical analysis of the dataset, the authors used polarized words in text, 'smile' and 'look away' in visual, and pauses and pitch in aural modality, as the main features (Tables 3.3 and 3.4).

3.2.1.2 MOUD Dataset

The Multimodal Opinion Utterances Dataset (MOUD) was developed in 2013 by Perez-Rosas et al. [273]. This is a dataset of utterances, with all videos recorded in Spanish. A final set of 80 videos was selected, out of which 65 were from

Table 3.3 State of the art of multimodal affect recognition where the text modality has been used

References	Data type	Modality	Fusion type
Chuang and Wu (2004) [67]	act	A+T	dec
Forbes-Riley and Litman (2004) [118]	nat	A+T	feat
Litman and Forbes-Riley (2004) [212]	nat	A+T	feat
Rigoll et al. (2005) [322]	act	A+T	dec
Litman and Forbes-Riley (2006) [213]	nat	A+T	feat
Seppi et al. (2008) [340]	ind	A+T	feat
Eyben et al. (2010) [110]	ind	A+T	model
Schuller (2011) [338]	nat	A+T	feat
Wu and Liang (2011) [400]	act	A+T	dec
Rozgic et al. (2012) [326]	act	A + T + V	feat
Savran et al. (2012) [333]	ind	A + T + V	model
Rosas et al. (2013) [324]	nat	A + T + V	feat
Wollmer et al. (2013) [398]	nat	A + T + V	hybrid
Sarkar et al. (2014) [332]	nat	A + T + V	feat
Alam et al. (2014) [2]	nat	A + T + V	dec
Ellis et al. (2014) [107]	nat	A + T + V	dec
Poria et al. (2014) [2]	act	A + T + V	feat
Siddiquie et al. (2015) [356]	nat	A + T + V	hybrid
Cai et al. (2015) [35]	nat	T+V	dec
Ji et al. (2015) [164]	nat	T+V	model
Yamasaki et al. (2015) [408]	nat	A+T	model
Poria et al. (2016) [294]	nat	A + T + V	feat

Legenda: Data Type (*act* Acted, *ind* Induced, *nat* Natural); Modality (*V* Video, *A* Audio, *T* Text); Fusion Type (*feat* Feature, *dec* Decision)

female speakers and 15 from male speakers, with age ranging from 20 to 60 years. A multimodal dataset of 498 utterances was eventually created with an average duration of 5 s and a standard deviation of 1.2 s. The dataset was annotated using Elan, an annotator tool used for video and audio sources along with two other annotators. The annotation task led to 182 positive, 231 negative and 85 neutral labeled utterances. There were 28 features considered for computation in total, including: prosody features, energy features, voice probabilities and spectral features. This trimodality dataset is said to produce an error rate reduction of 10.5% compared to the best unimodality. The authors also experimentally demonstrated an interesting fact that, a 'distressed brow' is the strongest feature for segment classification, with smile coming out a close second.

Table 3.4 State of the art of visual-audio affect recognition

References	Data type	Modality	Fusion type
Busso et al. (2004) [34]	act	V + A	feat
Chen et al. (2005) [62]	act	V + A	feat
Gunes and Piccardi (2005) [132]	act	V + B	feat
Hoch et al. (2005) [146]	act	V + A	dec
Kapoor and Picard (2005) [179]	nat	V + B + C	model
Kim et al. (2005) [187]	ind	A + Pp	feat
Wang and Guan (2005) [386]	act	V + A	feat
Zeng et al. (2005) [422]	act	V + A	model
Gunes and Piccardi (2005) [131]	act	V + B	feat
Pal et al. (2006) [258]	nat	V + A	dec
Sebe et al. (2006) [339]	act	V + A	model
Zeng et al. (2006) [419]	nat	V + A	model
Caridakis et al. (2006) [52]	ind	V + A + B	model
D'Mello and Graesser (2007) [90]	nat	B + C	feat
Gong et al. (2007) [128]	act	V + B	feat
Han et al. (2007) [136]	act	V + A	dec
Joo et al. (2007) [173]	act	V + A	dec
Karpouzis et al. [182] (2007)	ind	V + A	feat
Kim (2007) [187]	ind	A + Pp	feat
Schuller et al. (2007) [337]	nat	V+A	feat
Shan et al. (2007) [354]	act	V + B	feat
Zeng et al. (2007) [423]	act	V + A	model
Haq et al. (2008) [139]	act	V + A	feat
Kanluan et al. (2008) [178]	nat	V + A	dec
Metallinou et al. (2008) [229]	act	V + A	dec
Wang and Guan (2008) [386]	act	V + A	feat
Wimmer et al. (2008) [393]	ind	V + A	feat
Bailenson et al. (2008) [12]	ind	V + Pp	feat
Castellano et al. (2008)[53]	act	V + A + B	feat
Chetty and Wagner (2008) [66]	act	V + A	hybrid
Castellano et al. (2009) [53]	nat	V + C	feat
Emerich et al. (2009) [108]	act	V + A	feat
Gunes and Piccardi (2009) [133]	act	V + B	feat
Haq and Jackson (2009) [138]	act	V+A	dec
Khalali and Moradi (2009) [186]	ind	Cp + Pp	feat
Paleari et al. (2009) [259]	act	V + A	model
Rabie et al. (2009) [311]	act	V + A	model
D'Mello and Graesser (2010) [91]	nat	V + A + C	feat
Dy et al. (2010) [99]	nat	V + A	dec
Gajsek et al. (2010) [120]	act	V + A	dec
Kessous et al. (2010) [185]	act	V + A + B	feat
Kim and Lingenfelser (2010) [188]	ind	A + Pp	dec

(continued)

Table 3.4 (continued)

Mansoorizadeh and Charkari (2010) [222]	act	V+A	hybrid
Mansoorizadeh and Charkari (2010) [222]	act	V+A	hybrid
Wollmer et al. (2010) [397]	act	V + A	feat
Glodek et al. (2011) [125]	ind	V + A	dec
Banda and Robinson (2011) [18]	act	V + A	dec
Chanel et al. (2011) [55]	nat	Cp + Pp	dec
Cueva et al. (2011) [76]	act	V + A	dec
Datcu and Rothkrantz (2011) [83]	act	V + A	feat
Jiang et al. (2011) [166]	act	V+A	model
Lingenfelser et al. (2011) [211]	act	V + A	dec
Lingenfelser et al. (2011) [211]	ind	V + A	dec
Nicolaou et al. (2011) [249]	ind	V + A + B	model
Vu et al. (2011) [381]	act	A + B	dec
Wagner et al. (2011) [382]	act	V + A + B	dec
Walter et al. (2011) [383]	ind	A + Pp	dec
Hussain et al. (2012) [158]	ind	V + Pp	dec
Koelstra et al. (2012) [192]	ind	Cp + C + Pp	dec
Lin et al. (2012) [208]	act	V + A	model
Lin et al. (2012) [208]	ind	V + A	model
Lu and Jia (2012) [217]	act	V + A	model
Metallinou et al. (2012) [230]	act	V + A	model
Monkaresi et al. (2012) [238]	ind	V + Pp	feat
Park et al. (2012) [265]	act	V + A	dec
Rashid et al. (2012) [315]	act	V + A	dec
Soleymani et al. (2012) [362]	ind	Cp + Gaze	dec
Tu et al. (2012) [372]	act	V + A	dec
Baltrusaitis et al. (2013) [15]	ind	V + A	model
Dobrisek et al. (2013) [93]	act	V + A	dec
Glodek et al. (2013) [124]	ind	V + A	dec
Hommel et al. (2013) [147]	act	V + A	dec
Krell et al. (2013) [194]	ind	V + A	dec
Wollmer et al. (2013a) [396]	ind	V + A	model
Chen et al. (2014) [64]	act	V + A	dec
Wang et al. (2014) [385]	ind	Cp + C	feat

Legenda: Data Type (*act* Acted, *ind* Induced, *nat* Natural); Modality (*V* Video, *A* Audio, *B* Body, *Pp* Peripheral physiology, *CP* Central physiology, *Content* Content/context); Fusion Type (*feat* Feature, *dec* Decision)

3.2.1.3 ICT-MMMO Database

The Institute for Creative Technologies Multi-Modal Movie Opinion (ICT-MMMO) database was developed in 2013 by Wollmer et al. [398]. This dataset is a collection of online videos obtained from YouTube and ExpoTV reviewing movies in English. The authors used keywords such as movie review videos and opinions and the names of the recent movies as listed by imdb.com, as search keywords. The authors collected 308 YouTube videos, out of which 228 were annotated as positive, 57 as negative and 23 as neutral. They also gathered 78 movie review videos from ExpoTV, from which 62 were annotated as negative, 14 as neutral and 2 as positive. The final dataset comprised a total of 370 videos, which included all 308 videos from YouTube and 62 negative movie review videos from ExpoTV. The annotation task was performed by two annotators for YouTube videos and one annotator for ExpoTV videos. In contrast with other datasets, this dataset had five sentiment labels: strongly positive, weakly positive, neutral, strongly negative and weakly negative.

3.2.2 Datasets for Multimodal Emotion Recognition

We describe the datasets currently available for multimodal emotion recognition below. To the best of our knowledge, all available datasets for multimodal emotion recognition are acted.

3.2.2.1 HUMAINE Database

This dataset was developed in 2007 by Douglas-Cowie et al. [97]. The database provides naturalistic clips of pervasive emotions from multiple modalities and labels the best ones describing them. It consists of 50 clips from both naturalistic and induced data, spanning a broad emotional space, covering cues from body gestures, face, voice and words and representing speakers from different genders and cultures. Labels describing both signs and emotional content are designed to be time aligned rather than global, as timing appears to be an important factor in many areas.

3.2.2.2 The Belfast Database

This dataset was developed in 2000 by Douglas-Cowie et al. [96]. The database consists of audiovisual data of people discussing emotional subjects and are taken from TV chat shows and religious programs. It comprises 100 speakers and 239 clips, with 1 neutral and 1 emotional clip for each speaker. Two types of descriptors were provided for each clip – dimensional and categorical. Activation and evaluation are dimensions that are known to discriminate effectively between emotional states.

Activation values indicate the dynamics of a state and evaluation values provide a global indication of the positive or negative feeling associated with the emotional state. Categorical labels describe the emotional content of each state.

3.2.2.3 The SEMAINE Database

This dataset was developed in 2007 by McKeown et al. [226]. It is a large audiovisual database created for building agents that can engage a person in a sustained and emotional conversation using Sensitive Artificial Listener (SAL) [95] paradigm. SAL is an interaction involving two parties: a 'human' and an 'operator' (either machine or a person simulating a machine). The interaction is based on two qualities: one is low sensitivity to preceding verbal context (the words the user used that do not dictate whether to continue the conversation) and the second is conduciveness (response to a phrase by continuing the conversation). There were 150 participants, 959 conversations, each lasting 5 min. There were 6–8 annotators per clip, who eventually traced 5 affective dimensions and 27 associated categories. For the recordings, the participants were asked to talk in turn to four emotionally stereotyped characters. The characters are Prudence, who is even-tempered and sensible; Poppy, who is happy and outgoing; Spike, who is angry and confrontational; and Obadiah, who is sad and depressive. Videos were recorded at 49.979 frames per second at a spatial resolution of 780×580 pixels and 8 bits per sample, while audio was recorded at 48 kHz with 24 bits per sample. To accommodate research in audio-visual fusion, the audio and video signals were synchronized with an accuracy of 25 micro-seconds.

3.2.2.4 Interactive Emotional Dyadic Motion Capture Database (IEMOCAP)

IEMOCAP dataset was developed in 2008 by Busso et al. [33]. Ten actors were asked to record their facial expressions in front of cameras. Facial markers, head and hand gesture trackers were placed in order to collect the facial expressions, head and hand gestures. In particular, the dataset contains a total of 10 h recording of dyadic sessions, each of them expressing one of the following emotions: happiness, anger, sadness, frustration and neutral state. The recorded dyadic sessions were later manually segmented at utterance level (defined as continuous segments when one of the actors was actively speaking). The acting was based on some scripts, hence, it was easy to segment the dialogs for utterance detection in the textual part of the recordings. Busso et al. [33] used two famous emotion taxonomies in order to manually label the dataset at utterance level: discrete categorical-based annotations (i.e., labels such as happiness, anger, and sadness), and continuous attribute-based annotations (i.e., activation, valence and dominance). To assess the emotion categories of the recordings, six human evaluators were appointed. Having two different annotation schemes can provide complementary information in human-machine interaction systems. The evaluation sessions were organized so that three different evaluators assessed each utterance. Self-assessment manikins (SAMs) were also employed to evaluate the corpus in terms of the attributes valence [1-negative, 5-positive], activation [1-calm, 5-excited], and dominance [1-weak, 5-strong]. Two more human evaluators were asked to estimate the emotional content in recordings using the SAM system. These two types of emotional descriptors facilitate the complementary insights about the emotional expressions of humans, emotional communications between people which can further help develop better human-machine interfaces by automatically recognizing and synthesizing emotional cues expressed by humans.

3.2.2.5 The eNTERFACE Database

This dataset was developed in 2006 by Martin et al. [223]. It is an audiovisual developed for use as a reference database for testing and evaluating video, audio or joint audio-visual emotion recognition algorithms. This database elicited universal emotions of happiness, sadness, surprise, anger, disgust and fear with the help of 42 speakers, from 14 different nationalities.

3.2.3 Affective Detection from Textual Modality

In this section, we present the state of the art of both emotion recognition and sentiment analysis from text. The task of automatically identifying fine-grained emotions, such as anger, joy, surprise, fear, disgust, and sadness, explicitly or implicitly expressed in text has been addressed by several researchers [3, 365]. So far, approaches to text-based emotion and sentiment recognition rely mainly on rule-based techniques, bag of words (BoW) modeling using a large sentiment or emotion lexicon [235], or statistical approaches that assume the availability of a large dataset annotated with polarity or emotion labels [251].

Several supervised and unsupervised classifiers have been built to recognize emotional content in text [409]. The SNoW architecture [60] is one of the most useful frameworks for text-based emotion recognition. In the past decade, researchers have mainly focused on sentiment extraction from texts of different genres, such as news [109], blogs [209], Twitter messages [254], and customer reviews [149] to name a few.

Sentiment analysis systems can be broadly categorized into knowledge-based or statistics-based systems [36]. While initially the use of knowledge bases was more popular for the identification of emotions and polarity in text, recently sentiment analysis researchers have been increasingly using statistics-based approaches, with a special focus on supervised statistical methods. For example, Pang et al. [262] compared the performance of different machine learning algorithms on a movie review dataset and obtained 82.90% accuracy using a large number of textual features. A recent approach by Socher et al. [360] obtained even better accuracy (85%) on the same dataset using a recursive neural tensor network (RNTN). Yu and Hatzivassiloglou [414] used semantic orientation of words to identify polarity at sentence level. Melville et al. [228] developed a framework that exploits word-class association information for domain-dependent sentiment analysis.

Other unsupervised or knowledge-based approaches to sentiment analysis include: Turney et al. [373], who used seed words to calculate polarity and semantic orientation of phrases; Melville et al. [150], who proposed a mathematical model to extract emotional clues from blogs and used them for sentiment recognition; Gangemi et al. [122], who presented an unsupervised frame-based approach to identify opinion holders and topics.

3.2.3.1 Single- vs. Cross-Domain

Sentiment analysis research can also be categorized as single-domain [177, 262, 310, 373] versus cross-domain [26]. The work presented in [260] discusses spectral feature alignment to group domain-specific words from different domains into clusters. Authors first incorporated domain-independent words to aid the clustering process and then exploited the resulting clusters to reduce the gap between domain-specific words of two domains.

Bollegala et al. [28] developed a sentiment-sensitive distributional thesaurus by using labeled training data from a source domain and unlabeled training data from both source and target domains. Sentiment sensitivity was obtained by including documents' sentiment labels into the context vector. At the time of training and testing, this sentiment thesaurus was used to expand the feature vector.

Some recent approaches used SenticNet [45], a domain-independent resource for sentiment analysis containing 50,000 commonsense concepts, for tasks such as opinion holder detection [122], knowledge expansion [402], subjectivity detection [65], event summarization [349], short text message classification [123], sarcasm detection [287], Twitter sentiment classification [29], deception detection [162], user profiling [404], emotion visualization [334], and business intelligence [100].

3.2.3.2 Use of Linguistic Patterns

Whilst machine learning methods for supervised training of the sentiment analysis system are predominant in literature, a number of unsupervised methods such as linguistic patterns [255–257, 297, 301] can also be found. In theory, sentence structure is key to carry out sentiment analysis, as a simple change in the word order can flip the polarity of a sentence. Linguistic patterns aim to better understand sentence structure based on its lexical dependency tree, which can be used to calculate sentiment polarity.

One of the earliest works in the study of linguistic patterns for sentiment analysis was carried out by [141], where a corpus and some seed adjective sentiment words were used to find additional sentiment adjectives in the corpus. Their technique exploited a set of linguistic rules on connectives (‘and’, ‘or’, ‘but’, ‘either/or’, ‘neither/nor’) to identify more sentiment words and their orientations. In this way, they defined the idea of *sentiment consistency*.

Kanayama et al. [177] extended the approach by introducing definitions of intra-sentential (within a sentence) and inter-sentential (between neighboring sentences) sentiment consistency. Negation plays a major role in detecting the polarity of sentences. In [165], Jia et al. carried out an experiment to identify negations in text using linguistic clues and showed a significant performance improvement over the state of the art. However, when negations are implicit, e.g., cannot be recognized by an explicit negation identifier, sarcasm detection needs to be considered as well.

In [320], three conceptual layers, each of which consists of 8 textual features, was proposed to grasp implicit negations. A method to exploit discourse relations for the detection of tweets polarity was proposed in [61]. The authors showed how conjunctions, connectives, modals and conditionals might affect sentiments in tweets. In [280], a discourse parser combined information of sentential syntax, semantics and lexical information to build a tree that served as a representation of the discourse structure.

Wolf et al. [395] presented a method to represent discourse coherence, using contentful conjunctions to illustrate coherence relations. Discourse coherence relations have also been explored in [389] and, in [314], discourse connectives identification is applied to biomedical text. Liu et al. [215] proposed a collection of opinion rules implying positive or negative sentiment. First, the rules at conceptual level are described without considering how they may be expressed in actual sentences, i.e., without considering context. Next, an inspection at expression level combines more than one input-constituent expression to derive an overall sentiment orientation for the composite expression.

Moilanen et al. [237] introduced the notion of sentiment conflict, which is used when opposite sentiment words occur together, e.g., ‘terribly good’. Conflict resolution is achieved by ranking the constituents on the basis of relative weights assigned to them, considering which constituent is more important with respect to sentiment. In [89], a holistic lexicon-based approach was used to evaluate the semantic orientations of opinions expressed on product features in reviews, by exploiting external evidence and linguistic conventions of natural language expressions. This approach, implemented in a system called Opinion Observer, allows for handling opinion words that are context-dependent. The authors found that both aspect and sentiment expressing words are important and proposed using the pair (*aspect, sentiment word*) as an *opinion context*.

3.2.3.3 Bag of Words Versus Bag of Concepts

Text representation is a key task for any text classification framework. BoW looks for surface word forms and does not consider semantic and contextual clues in text. Most of the well-known techniques have focused on BoW representation for text classification [174, 190, 246]. To overcome the problem of limited capability in grasping semantic clues, some existing related works relied on using knowledge bases [205, 275].

The bag of concepts (BoC) model leverages on representing text as a conceptual vector rather than relying on terms in text. For example, if a text contains "red" and "orange" then, BoC models them as the concept "color", e.g., BoC looks for hyponym. The BoC model was first proposed by Sahlgren et al. [330] to enhance the performance of SVM in text categorization tasks. According to their method, concepts are synonym sets of BoW. Among recent approaches adopting the BoC model, Wang et al. [384] presented the idea of concept as a set of entities in a given domain, e.g., words belonging to similar classes have similar representation.

If a document contains "Jeep" and "Honda" then both of these words can be conceptualized by "Car". On the basis of their study, we identify two major advantages of the BoC model:

- *Replacement of surface matching with semantic similarity:* the BoC model calculates semantic similarity between words and multi-word expressions at a higher level.
- *Tolerance with new terms:* once concepts related to a category are modeled, BoC is able to handle new words under that category.

In [426], Zhang et al. discussed semantic classification on a disease corpus. Though their approach does not focus on the BoC model, they attempted to capture semantic information from text at a higher level. According to their study, use of contextual semantic features along with the BoW model can be very useful for semantic text classification. Wu et al. [399] built a sentiment lexicon using a commonsense knowledge base. Under the hypothesis that concepts pass their sentiment intensity to neighbors based on the relations connecting them, they constructed an enriched sentiment lexicon able to perform better on sentiment polarity classification tasks.

Concept-based approaches to sentiment analysis focus on a semantic analysis of text through the use of web ontologies or semantic networks, which allow the aggregation of conceptual and affective information associated with natural language opinions. By relying on large semantic knowledge bases, such approaches step away from the blind use of keywords or word co-occurrence counts and, instead, rely on implicit features associated with natural language concepts [50]. Unlike syntactical techniques, concept-based approaches are also able to detect sentiments that are expressed in a subtle manner, e.g., through the analysis of concepts that do not explicitly convey any emotion, but are implicitly linked to other concepts that do so.

The analysis at concept level is intended to infer the semantic and affective information associated with natural language opinions and, hence, to enable a comparative fine-grained aspect-based sentiment analysis. Rather than gathering isolated opinions about a whole item (e.g., iPhone7), users are generally more interested in comparing different products according to their specific features (e.g., iPhone7's vs Galaxy S7's touchscreen), or even sub-features (e.g., fragility of iPhone7's vs Galaxy S7's touchscreen). In this context, the construction of comprehensive common and commonsense knowledge bases is key for aspect extraction and polarity detection, respectively. Commonsense, in particular, is necessary to appropriately deconstruct natural language text into sentiments; for example, to appraise the concept 'small room' as negative and 'small queue' as positive, or the concept 'go read the book' as positive for a book review but negative for a movie review.

3.2.3.4 Contextual Subjectivity

Wilson et al. [392] reported that, although a word or phrase in a lexicon is marked positive or negative, in the context of the sentence expression it may have no sentiment or even have opposite sentiment.

In their work, subjective expressions were first labeled, with the goal of the work aimed at classifying the contextual sentiment of the given expressions. The authors employed a supervised learning approach based on two steps: first, it determined whether the expression is subjective or objective, second it determined whether the subjective expression is positive, negative, or neutral.

In [8], authors presented an analysis of opinions based on a lexical semantic analysis of a wide class of expressions coupled together, inspecting how clauses involving these expressions were related to each other within a discourse. Narayanan et al. [243] aimed to analyze the sentiment polarity of conditional sentences, studying the linguistic structure of such sentences and applying supervised learning models for dynamic classification.

3.2.3.5 New Era of NLP: Emergence of Deep Learning

Deep-learning architectures and algorithms have already made impressive advances in fields such as computer vision and pattern recognition [38, 46, 283, 284, 286, 290, 291, 293, 299]. Following this trend, recent NLP research is now increasingly focusing on the use of new deep learning methods. As demonstrated in [71], a simple deep learning framework outperforms most state-of-the-art approaches, in several NLP tasks such as named entity recognition (NER), sequential role labeling (SRL), and part of speech (POS) tagging.

Alternative approaches have exploited the fact that many short n-grams are neutral while longer phrases are well distributed among positive and negative subjective sentence classes. Thus, matrix representations for long phrases and matrix multiplication to model composition are also being used to evaluate sentiment.

In such models, sentence composition is modeled using deep neural networks such as recursive auto-associated memories [126, 176]. Recursive neural networks predict the sentiment class at each node in the parse tree and attempt to capture

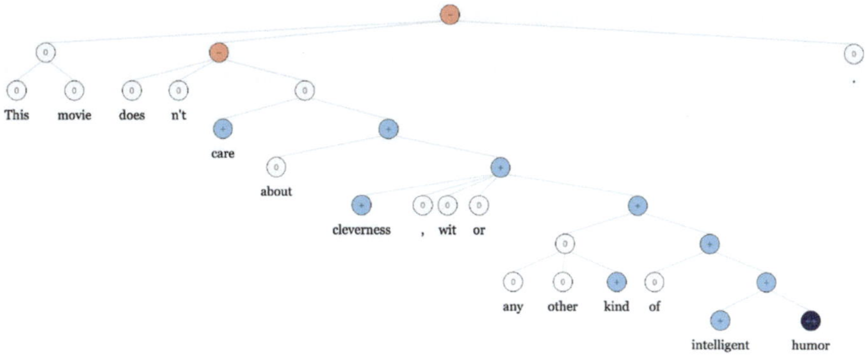

Fig. 3.1 RNTN applied on the dependency tree of the sentence "This movie doesn't care about cleverness, wit or any other kind of intelligent humor"

the negation and its scope in the entire sentence. In the standard configuration, each word is represented as a vector and it is first determined which parent already has all its children computed. Next, the parent is computed via a composition function over child nodes, which depends on words being combined and, hence, is linguistically motivated. However, the number of possible composition functions is exponential, hence, in [360], a RNTN was introduced (Fig. 3.1), which uses a single tensor composition function to define multiple bilinear dependencies between words.

More recently, a new trend has emerged [189] focusing on the use of word embeddings pre-trained on a large corpus [231]. In such methods, word vectors are typically concatenated to form a sentence or document vector and then fed to a deep network for training. Studies show that these methods outperform state-of-the-art feature extraction based opinion mining methods, thus establishing themselves as new state-of-the-art benchmarks [189].

3.3 Visual, Audio Features for Affect Recognition

3.3.1 Visual Modality

Facial expressions are primary cues for understanding emotions and sentiments. Across the ages of people involved, and the nature of conversations, facial expressions are the primary channel to form an impression of the subject's present state of mind. Ekman et al. [104], considered the pioneers in this research, argued that it is possible to detect six basic emotions, e.g., Anger, Joy, Sadness, Disgust and Surprise from cues of facial expressions. In this section, we present various studies on the use of visual features for multimodal affect analysis.

3.3.1.1 Facial Action Coding System

As facial cues gained traction in discerning emotions, a number of observer-based measurement systems for facial expressions were developed [160, 195, 269]. Out of these systems, the Facial Action Coding System (FACS) developed by Ekman and Friesen [269] has been widely used. FACS is based on the reconstruction of facial expressions in terms of Action Units (AU). The facial muscles of all humans are almost identical and AUs are based on movements of these facial muscles, which consist of three basic parts: AU number, FACS name, and muscular basis. FACS only distinguishes facial actions and gives no inference on emotions. FACS codes are used to infer emotions using a variety of available resources such as FACS Investigators' Guide [103], the FACS interpretive database [102], and a large body of empirical research [105].

These resources use combinations of AUs for specifying emotions. In 1992, the seventh emotion 'contempt' was added to the universal set of emotions [224], as it expresses *disrespect* which is equally important when compared to the six basic emotions. In 2002, an updated version of FACS was introduced where the description of each AU, and AU combinations were refined. Furthermore, details on head movements and eye positions were also added [103]. In addition to emotion and sentiment analysis, FACS is also used in the field of neuroscience [323], computer vision [19], computer graphics [30] and animation [266], and face encoding for digital signal processing [369].

Ekman's work inspired many researchers to employ image and video processing methods in order to analyze facial expressions. Yacoob et al. [407] and Black et al. [23] used high gradient points on face, track head and facial movements to recognize facial expressions. Geometrical features [427] with a multi-scale, multi-orientation Gabor Wavelet-based representation was used to identify expressions. Kalman Filter and probabilistic principal component analysis (PCA) [140] was used to track the pupils, in order to enhance the features. A stochastic gradient descent based technique [172] and active appearance model (AAM) [72] were used to recover the face shape and texture parameters, for facial features.

A comparison of several techniques [94], such as optical flow, PCA, independent component analysis (ICA), local feature analysis and Gabor wavelet, for recognition of action units, found that, Gabor wavelet representation and ICA performed better on most datasets. Considering every part of the face as an important feature, a multi-state face component model [370], was introduced to exploit both permanent and transient features. Permanent features are those features that remain the same through ages, which include opening and closing of lips and eyes, pupil location, eyebrows and cheek areas.

Transient features are observed only at the time of facial expressions, such as contraction of the corrugator muscle that produces vertical furrows between the eyebrows. Texture features of the face have also been considered for facial expression analysis in a number of feature extraction methods, including: image intensity [19], image difference [114], edge detection [370], and Gabor wavelets [219]. In order to recognize and model facial expressions in terms of emotions and sentiments, numerous classifiers have been used, such as Nearest Neighbor [114], Neural Networks [370], support vector machine (SVM) [214], Bayesian Networks [69], and AdaBoost classifiers [387].

3.3.1.2 Main Facial Expression Recognition Techniques

Some of the important facial expression recognition techniques, face tracking and feature extraction methods are briefly described below:

Active Appearance Models (AAM) [199] are well-known algorithms for modeling deformable objects. The models decouple the shape and texture of objects, using a gradient-based model fitting approach. Most popular applications of AAM include recognition, tracking, segmentation and synthesis.

Optical flow models [407] are used to calculate the motion of the objects or the motion of two image frames, based on gradients. These methods are also called differential methods as they are calculated using Taylor series.

Active Shape Models (ASM) [73] are statistical models that deform to fit the data or object in an image in ways consistent with the training data provided. These models are used mainly to enhance automatic analysis of images under noisy or cluttered environments.

3D Morphable Models (3DMM) [24] are models that are used for facial feature analysis by modeling 3D faces, that are immune to pose and illumination. Thus, these models are used for automatic 3D face registration by computing dense one-to-one correspondences, and adjust the naturalness of modeled faces.

Muscle-based models [250] are models that consist of facial feature points corresponding to facial muscles, for tracking motion of facial components, such as eyebrows, eyes and mouth, thus recognizing facial expressions.

3D wireframe models [70] are 3-dimensional models of an object where the edges or vertices are connected using straight lines or curves. Once the model is designed for a given face, the head motion and local deformations of the facial features such as eyebrows, eyes and mouth can be tracked.

Elastic net model [191] represents facial expressions as motion vectors of the deformed net, from a facial edge image.

Geometry-based shape models [376] are models that represent expression changes in a face through geometry-based high-dimensional 2D shape transformations, which are then used to register regions of a face with expressions, to those defined on the template face.

3D Constrained Local Model (CLM-Z) [17] is a non-rigid face tracking model used to track facial features under varying poses, by including both depth and intensity information. Non-rigid face tracking refers to points of interest in an image, for example, nose tip, corners of eyes and lips. The CLM-Z model can be described by the parameters p = [s, R, q, t], where s is a scale factor, R is object rotation, t represents 2D translation and q is the vector describing non-rigid variation of the q.

Generalized Adaptive View-based Appearance Model (GAVAM) [240] is a probabilistic framework that combines dynamic or motion-based approaches to track the position and orientation of the head through video sequences, and employs static user-independent approaches to detect head pose from the image. GAVAM is considered a high-precision, user-independent real-time head pose tracking algorithm.

In other works, the CLM-Z and GAVAM models were integrated for rigid and non-rigid facial tracking to improve pose estimation accuracy for both 2D and 3D cases [17].

3.3.1.3 Extracting Temporal Features from Videos

Having addressed some of the key works on recognizing facial expressions from images, we would most of those methods do not work well for videos as they miss modeling the temporal information. In this paragraph, we discuss a few methods which used temporal information [57, 206, 214, 410], Motion-Units (MU) (otherwise called facial motion) [70] and features in terms of duration, content and valence [196] for affect recognition from videos.

An important facet in video-based methods is to maintain accurate tracking throughout the video sequence. A wide range of deformable models, such as muscle-based models [250], 3D wireframe model [70], elastic net model [191] and geometry-based shape model [84, 376], have been used to track facial features in videos. Thus, deformable models have been proven to improve both facial tracking and facial expression analysis accuracy [390]. Thus, many automatic methods for detection of facial features and face expressions were proposed [115, 263, 264], both image-based and video-based.

3.3.1.4 Body Gestures

Though most research works have concentrated on facial feature extraction for emotion and sentiment analysis, there are some contributions based on features extracted from body gestures. Research in psychology suggests that body gestures provide a significant source of features for emotion and sentiment recognition. In [85], a detailed study was carried out on how body gestures are related to emotions and how various combinations of body gesture dimensions and qualities can be found in different emotions. It was also shown how basic emotions can be automatically distinguished from simple statistical measures of motion's dynamics induced by body movements [180].

Based on these groundbreaking studies, a set of body gesture features for emotion recognition were extracted to help autistic children [276]. Inspired by these pioneering findings, an automatic emotion recognition framework was proposed from body gestures, using a set of postural, kinematic, and geometrical features extracted from sequences of 3D skeletal movements, which were fed to a multiclass SVM classifier for emotion recognition [277].

In [51], a mathematical model was developed to analyze the dynamics of body gestures for emotion expressiveness. Some of the extracted motion cues to understand the subject's temporal profile included: initial and final slope of the main peak, ratio between the maximum value and the duration of the main peak, ratio between the absolute maximum and the biggest following relative maximum, centroid of the energy, symmetry index, shift index of the main peak, and number of peaks.

In [14], both facial and hand gesture features were used to perform emotion analysis and the creation of moving skin masks in order to estimate user's movement by tracking the centroid of skin masks over the person under experimentation.

3.3.1.5 New Era: Deep Learning to Extract Visual Features

In the last two sections, we described the use of handcrafted feature extraction from visual modality and mathematical models for facial expression analysis. With the advent of deep learning, we can now extract features automatically without prior intervention. The deep learning framework enables robust and accurate feature learning, which in turn produces benchmark performance on a range of applications, including digit recognition [143], image classification [197], feature learning [202], visual recognition [184], musical signal processing [134] and NLP [58]. Both academia and industries have invested a huge amount of effort in building powerful deep neural networks.

These demonstrate the potential of deep learning for learning robust features, in both supervised and unsupervised settings. Even though deep neural networks may be trapped in local optima [142], different optimization techniques can be effectively employed to enhance their performance in many challenging fields. Inspired by the recent success of deep learning, emotion and sentiment analysis tasks have also been enhanced by the adoption of deep learning algorithms, e.g., convolutional neural network (CNN).

In [406], a novel visual sentiment prediction framework was designed to understand images using CNN. The framework is based on transfer learning from a CNN pre-trained on large scale data for object recognition, which in turn is used for sentiment prediction. The main advantage of the proposed framework is that there is no requirement of domain knowledge for visual sentiment prediction.

Motivated by the need for processing increasingly large and noisy data in the field of image sentiment analysis, CNN has been employed in [412], coupled with a progressive strategy to fine tune deep learning networks for filtering out noisy training data and with the use of domain transfer learning to enhance performance.

In [405], emotion recognition for user generated videos is performed through the extraction of deep convolution network features and through zero-shot emotion learning, a method that predicts emotions that are not observed in the training set. To implement this task, image transfer encoding (ITE) is proposed to encode the extracted features and generate video representations.

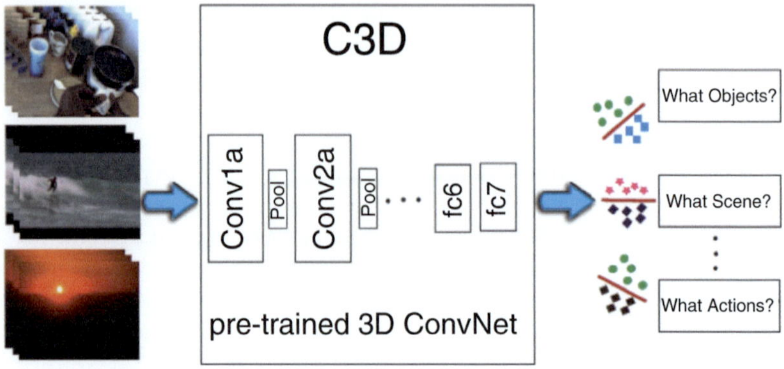

Fig. 3.2 C3D for extracting spatio-temporal generic video features

More recently, deep 3D convolutional networks (C3D) (Fig. 3.2) have been proposed for spatio-temporal feature learning [371]. The C3D network comprises 8 convolution layers, 5 pooling layers, 2 fully connected layers, and a softmax output layer. The network has been shown to be more amenable for spatio-temporal feature learning, in comparison with 2D convolution networks. The 3×3 convolution kernels in all layers were found to create the best performing architecture, with learned features using a simple classifier outperforming existing state-of-the-art methods.

3.3.2 Audio Modality

Similar to text and visual feature analysis, emotion and sentiment analysis through audio features has its specific components. Several prosodic and acoustic features have been used in the literature in order to teach machines to detect emotions [241, 242, 400, 401]. Since emotional characteristics are more prominent in the prosodic features, these features are widely used in the literature [193, 218].

Researchers started targeting affective reactions to everyday sounds [375], which have ultimately led to enormous applications to date, both in unimodal and multimodal analysis. The current trend is to understand affect in naturalistic videos [20, 144, 203], e.g., spontaneous dialogs, audio recordings collected in call centers, interviews, etc. Early research on extraction of audio features focused on phonetic and acoustic properties of spoken language. With the help of psychological studies related to emotion, it was found that vocal parameters, especially pitch, intensity, speaking rate and voice quality play an important role in recognition of emotion and sentiment analysis [242].

Further studies showed that acoustic parameters change not only through oral variations, but are also dependent on personality traits. Various works have been carried out based on the types of features that are needed for better analysis [88, 379]. Researchers have found out that pitch and energy related features as most important for affect recognition. Other features that have been used by some researchers for feature extraction include formants, mel frequency cepstral coefficients (MFCC), pause, teager energy operated based features, log frequency power coefficients (LFPC) and linear prediction cepstral coefficients (LPCC).

Some of the important audio features are described briefly below:

- *Mel Frequency Cepstral Coefficients (MFCC)* are coefficients that collectively form a mel-frequency cepstrum (MFC). The MFC is a short-term power spectrum of a sound or an audio clip, which approximates the human auditory system more closely than any other available linearly-spaced frequency band distribution. This feature is calculated based on the linear cosine transform of a log power spectrum, on a mel-frequency scaling.
- *Spectral centroid* indicates the center of mass of the magnitude spectrum, which simply provides an indication of the brightness of a sound.
- *Spectral flux* defines how quickly the power spectrum of a signal is changing. This feature is usually calculated by taking the Euclidean distance between two normalized spectra.
- *Beat Histogram* is a histogram showing the strength of different rhythmic periodicities in a signal. It is typically calculated by taking the RMS of 256 windows and then taking the FFT of the output.
- *Beat sum* is used to find regular beats in a signal. It is calculated as the sum of all entries in the beat histogram.
- *Strongest beat* is the strongest beat in a signal and is found by identifying the strongest bin the beat histogram.
- *Pause duration* is the time the speaker is silent in an audio segment.
- *Pitch* is the quality of a sound governed by the rate of vibrations producing it; the degree of highness or lowness of a tone.
- *The Perceptual Linear Predictive Coefficients (PLP)* are created from linear prediction coefficients (LPC) by performing perceptual processing before autoregressive modeling and followed by cepstral conversion.

OpenSMILE [111] is a popular audio feature extraction toolkit which is able to extract all the key features as elaborated above. This framework is shown in Fig. 3.3. The affective reactions to sound have been classified as discrete feeling states and states based on dimensions [204, 375]. Discrete feeling states are defined as emotions that are spontaneous, uncontrollable or, in other words, universal emotions. The states based on dimension are hedonic valence (pleasantness), arousal (activation, intensity) and dominance.

Recent studies on speech-based emotion analysis [82, 86, 170, 242] have focused on identifying several acoustic features such as fundamental frequency (pitch), intensity of utterance [63], bandwidth, and duration. The speaker-dependent approach often gives much better results than the speaker-independent approach, as shown by benchmark results of Navas et al. [244], where about 98% accuracy was achieved using the Gaussian mixture model (GMM) as a classifier, with prosodic, voice quality as well as MFCCs employed as speech features. However, the speaker-dependent approach is not feasible in many practical applications that deal with a very large number of users.

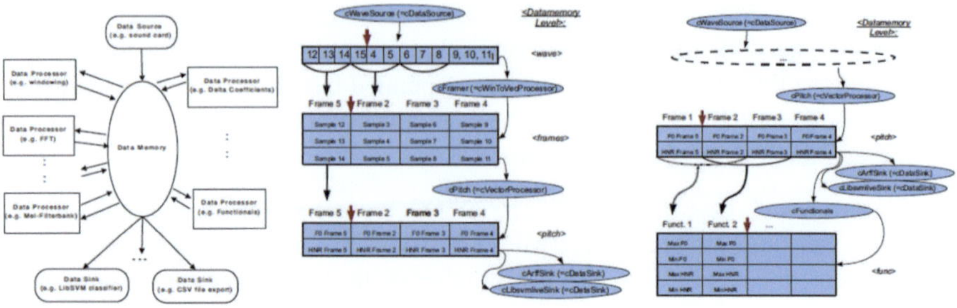

Fig. 3.3 The OpenSMILE framework

3.3.2.1 Local Features vs. Global Features

Audio affect classification is also classified into local features and global features. The common approach to analyze audio modality is to segment each audio or each utterance into either overlapped or non-overlapped segments and then examine them. Within a segment the signal is considered to be stationary. The features extracted from these segments are called local features.

In speech production, there are several utterances and, for each utterance, the audio signal can be divided into several segments. Global features are calculated by measuring several statistics, e.g., average, mean, deviation of the local features. Global features are the most commonly used features in the literature. They are fast to compute and, as they are fewer in number as compared to local features, the overall speed of computation is enhanced [106]. However, there are some drawbacks of calculating global features, as some of them are only useful to detect affect of high arousal, e.g., anger and disgust. For lower arousals, global features are not that effective, e.g., global features are less prominent to distinguish between anger and joy. Global features also lack temporal information and dependence between two segments in an utterance.

3.3.2.2 Speaker-Independent Applications

To the best of our knowledge, for speaker-independent applications, the best classification accuracy achieved so far is 81% [9], obtained on the Berlin Database of Emotional Speech (BDES) [32] using a two-step classification approach and a unique set of spectral, prosodic, and voice features, selected through the Sequential Floating Forward Selection (SFFS) algorithm [308]. As demonstrated in the analysis by Scherer et al. [335], human ability to recognize emotions from speech audio is about 60%. Their study showed that sadness and anger are detected more easily from speech, while the recognition of joy and fear is less reliable. Caridakis et al. [51] obtained 93.30% and 76.67% accuracy to identify anger and sadness, respectively, from speech, using 377 features based on intensity, pitch, MFCCs, Bark spectral bands, voiced segment characteristics, and pause length.

3.3.2.3 Audio Features Extraction Using Deep Networks

As for computer vision, deep learning is also gaining increasing attention in audio classification research. In the context of audio emotion classification, autoencoder followed by a CNN has been used in [157]. Authors trained CNN on the features extracted from all time frames. These types of models are usually incapable of modeling temporal information.

To overcome this problem, Long Short Term Memory (LSTM) [129], and bi-directional LSTM [110] have been commonly used on hand-extracted acoustic features. In computer vision, deep networks are frequently used for automatic feature extraction. A possible research question is whether deep networks can be replicated for automatic feature extraction from aural data. As shown in a pilot study [4], CNN can be used to extract features from audio, which can subsequently be used in a classifier for the final emotion classification task. Generalized discriminant analysis (GerDA) based deep neural networks are also a very popular approach in the literature for automatic feature extraction from raw audio data. However, most deep learning approaches in audio emotion classification literature rely on handcrafted features [135].

Recently, researchers have applied audio emotion and sentiment analysis in many fields, in particular to one of the most active and prominent areas in recent years: human-computer interaction [368, 380].

3.4 Multimodal Affect Recognition

Multimodal affect analysis has already created a lot of buzz in the field of affective computing. This field has now become equally important and popular among the computer scientists [13, 40, 48, 49, 59, 98, 281, 288, 289, 413, 416–418, 421]. In the previous section, we have discussed state-of-the-art methods which used either of the Visual, Audio or Text modalities for affect recognition. In this section, we discuss the approaches to solve the multimodal affect recognition problem.

3.4.1 Information Fusion Techniques

Multimodal affect recognition can be seen as the fusion of information coming from different modalities. Multimodal fusion is the process of combining data collected from various modalities for analysis tasks. It has gained increasing attention from researchers in diverse fields, owing to it potential for innumerable applications, including but not limited to: sentiment analysis, emotion recognition, semantic concept detection, event detection, human tracking, image segmentation, video classification, etc. The fusion of multimodal data can provide surplus information with an increase in accuracy [92] of the overall result or decision making. As the data collected from various modalities comes in various forms, it is also necessary to consider the period of multimodal fusion in different levels. To date, there are mainly two levels or types of fusion studied by researchers: feature-level fusion or early fusion, and decision-level fusion or late fusion. These have also been employed by some researchers as part of a hybrid fusion approach. Furthermore, there is ‘model-level fusion’, a type of multimodal fusion designed by researchers as per their application requirements.

Feature-level or early fusion [238, 324, 332, 385] fuses the features extracted from various modalities such as visual features, text features, audio features, etc., as a general feature vector and the combined features are sent for analysis. The advantage of feature-level fusion is that the correlation between various multimodal features at an early stage can potentially provide better task accomplishment. The disadvantage of this fusion process is time synchronization, as the features obtained belong to diverse modalities and can differ widely in many aspects, so before the fusion process takes place, the features must be brought into the same format.

3.4.1.1 Decision-Level or Late Fusion [2, 35, 93, 124, 408]

In this fusion process, the features of each modality are examined and classified independently and the results are fused as a decision vector to obtain the final decision. The advantage of decision-level fusion is that, the fusion of decisions obtained from various modalities becomes easy compared to feature-level fusion, since the decisions resulting from multiple modalities usually have the same form of data. Another advantage of this fusion process is that, every modality can utilize its best suitable classifier or model to learn its features. As different classifiers are used for the analysis task, the learning process of all these classifiers at the decision-level fusion stage, becomes tedious and time consuming. Our survey of fusion methods used to date has shown that, more recently, researchers have tended to prefer decision-level fusion over feature-level fusion. Among the notable decision-level fusion methods, Kalman filter has been in [124] as method to fuse classifiers. They considered video as a time dynamics or series and the prediction scores (between 0 and 1) of the base classifiers were fused using Kalman filter. On the other hand, Dobrivsek et al. [93] employed weight sum and weighted product rule for fusion. On the eNTERFACE dataset weighted product (accuracy: 77.20%) rule gave better result than weighted sum approach (accuracy: 75.90%).

3.4.1.2 Hybrid Multimodal Fusion [222, 398]

This type of fusion is the combination of both feature-level and decision-level fusion methods. In an attempt to exploit the advantages of both feature and decision-level fusion strategies and overcome the disadvantages of each, researchers opt for hybrid fusion. One such hybrid fusion proposed by Wollmer et al. [398] is shown in Fig. 3.4. As we can see in Fig. 3.4, in their method, audio and visual features were

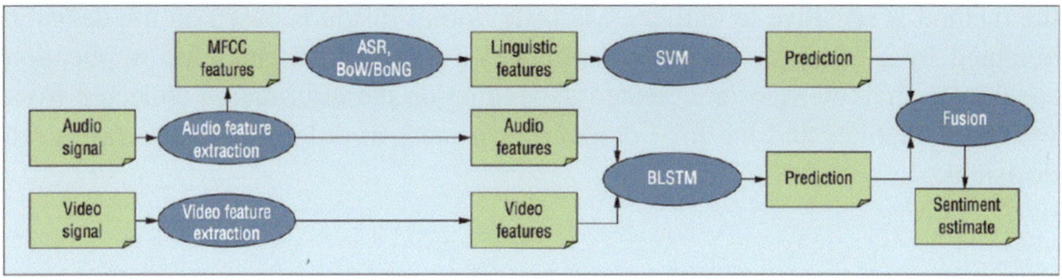

Fig. 3.4 Hybrid fusion for multimodal sentiment analysis in YouTube videos as proposed by [398]

fused using BLSTM at feature level. The result of that fusion were then fused with the prediction of the textual classifier using decision-level fusion.

3.4.1.3 Model-Level Fusion [15, 208, 217, 230, 396]

It is a technique that uses the correlation between data observed under different modalities, with a relaxed fusion of the data. Researchers built models satisfying their research needs and the problem space. Song et al. [363] used a tripled Hidden Markov Model (HMM) to model the correlation properties of three component HMM's based on audio-visual streams. Zeng et al. [420] proposed a Multi-stream Fused Hidden Markov Model (MFHMM) for audio-visual affect recognition. The MFHMM builds an optimal connection between various streams based on maximum entropy and maximum mutual information principle. Caridakis et al. [52] and Petridis et al. [274] proposed neural networks to combine audio and visual modalities for emotion recognition. Sebe et al. [339] proposed the Bayesian network topology to recognize emotions from audio-visual modalities, by combining the two modalities in a probabilistic manner. According to Atrey et al. [10], fusion can be classified into three categories: rule-based, classification-based and estimation-based methods. The categorization is based on the basic nature of the methods and the problem space, as outlined next.

3.4.1.4 Rule-Based Fusion Methods [75, 159]

As the name suggests, multimodal information is fused by statistical rule based methods such as linear weighted fusion, majority voting and custom-defined rules. The linear weighted fusion method uses sum or product operators to fuse features obtained from different modalities or decision obtained from a classifier. Before the fusion of multimodal information takes place, normalized weights are assigned to every modality under consideration. Thus, the linear weighted fusion method is computationally less expensive compared to other methods, however the weights need to be normalized appropriately for optimal execution. The drawback is that the method is sensitive to outliers. Majority voting fusion is based on the decision obtained by a majority of the classifiers. Custom-defined rules are application specific, in that, the rules are created depending on the information collected from various modalities and the final outcome expected, in order to achieve optimized decisions.

3.4.1.5 Classification-Based Fusion Methods [1, 245]

In this method, a range of classification algorithms are used to classify the multimodal information into pre-defined classes. Various methods used under this category include: SVMs, Bayesian inference, Dempster-Shafer theory, dynamic bayesian networks, neural networks and maximum entropy models. SVM is probably the most widely used supervised learning method for data classification tasks. In this method, input data vectors are classified into predefined learned classes, thus solving the pattern classification problem in view of multimodal fusion. The method is usually applicable for decision-level and hybrid fusion. The Bayesian inference fusion method fuses multimodal information based on rules of probability theory. In this method, the features from various modalities or the decisions obtained from various classifiers are combined and an implication of the joint probability is derived. The Dempster-Shafer evidence theory generalizes Bayesian theory of subjective probability. This theory allows union of classes and also represents both uncertainty and imprecision, through the definition of belief and plausibility functions. The Dempster-Shafer theory is a statistical method and is concerned with fusing independent sets of probability assignments to form a single class, thus relaxing the disadvantage of the Bayesian inference method. The Dynamic Bayesian Network (DBN) is an extension of the Bayesian inference method to a network of graphs, where the nodes represent different modalities and the edges denote their probabilistic dependencies. The DBN is termed by different names in the literature such as Probabilistic Generative Models, graphical models, etc. The advantage of this network over other methods is that the temporal dynamics of multimodal data can easily be integrated. The most popular form of DBN is the Hidden Markov Model (HMM). The Maximum entropy model is a statistical model classifier which follows an information-theoretic approach and provides probability of observed classes. Finally, the other widely used method is Neural Networks. A typical neural network model consists of input, hidden and output nodes or neurons. The input to the network can be features of different modality or decisions from various classifiers. The output provides fusion of data under consideration. The hidden layer of neurons provides activation functions to produce the expected output, and the number of hidden layers and neurons are chosen to obtain the desired accuracy of results. The connections between neurons have specific weights which can be appropriately tuned for the learning process of the neural network, to achieve the target performance accuracy.

3.4.1.6 Estimation-Based Fusion Methods [247, 305]

This category includes kalman filter, extended kalman filter and particle filter based fusion methods. These methods are usually employed to estimate the state of moving object using multimodal information, especially audio and video. The kalman filter is used for real-time dynamic, low-level data and provides state estimates for the system. This model does not require storage of the past of the object under observation, as the model only needs the state estimate of the previous time stamp. However the kalman filter model is restricted to linear systems. Thus, for systems with non-linear characteristics, extended kalman filter is used. Particle filters, also known as Sequential Monte Carlo model, is a simulation-based method used to obtain the state distribution of non-linear and non-Gaussian state-space models.

3.4.2 Recent Results

In this section, we describe recent key works in multimodal affect recognition. We summarize state-of-the-art methods, their results and categorize the works based on the datasets described in Sect. 3.2.

3.4.2.1 Multimodal Sentiment Analysis

MOUD Dataset

The work by Perez et al. [273] focused on multimodal sentiment analysis using the MOUD dataset based on visual, audio and textual modalities. FACS and AUs were used as visual features and openEAR was used for extracting acoustic, prosodic features. Simple unigrams were used for textual feature construction. The combination of these features were then fed to an SVM for fusion and 74.66% accuracy was obtained.

YouTube Dataset

Morency et al. [239] extracted facial features like smile detection and duration, look away and audio features like pause duration for sentiment analysis on the YouTube dataset. As textual features, two lexicons containing positive and negative words were developed from the MPQA corpus distribution. They fused and fed those features to a Hidden Markov Model (HMM) for final sentiment classification. However, the accuracy was relatively lower (55.33%). Possible future work would be to use more advanced classifiers, such as SVM, CNN, coupled with the use of complex features.

ICT-MMMO Dataset

Wollmer et al. [398] used the same mechanism as Morency et al. [239] for audio-visual feature extraction. In particular, OKAO vision was used to extract visual features which were then fed to CFS for feature selection. In the case of audio feature extraction, they used openEAR. Simple Bag-Of-Words were utilized as text features. Audio-visual features were fed to a Bidirectional-LSTM (BLSTM) for early feature-level fusion and SVM was used to obtain the class label of the textual modality. Finally, the output of BLSTM and SVM were fused at the decision level, using a weighted summing technique.

3.4.2.2 Multimodal Emotion Recognition

Recent Works on the SEMAINE Dataset

Gunes et al. [130] used visual aspects which aimed to predict dimensional emotions from spontaneous head gestures. Automatic detection of head nods and shakes was based on 2-dimensional (2D) global head motion estimation. In order to determine the magnitude and direction of the 2D head motion, optical flow was computed between two consecutive frames. It was applied to a refined region (i.e., resized and smoothed) within the detected facial area to exclude irrelevant background information. Directional codewords were generated by the visual feature extraction module, and fed into a HMM for training a nodHMM and a shakeHMM. A Support Vector Machine for Regression (SVR) was used for dimensional emotion prediction from head gestures. The final feature set was scaled in the range $[-1, +1]$. The parameters of SVR, for each coder-dimension combination, were optimized using 10-fold cross-validation for a subset of the data at hand. The MSE for detection of valence, arousal and other axis was found to be 0.1 on average, as opposed to 0.115 resulting from human annotators.

Valstar et al. [374] focussed on FACS Acton Units detection and intensity estimation, and derived its datasets from SEMAINE and BP4D-Spontaneous database. The training partition (SEMAINE database) consisted of 16 sessions, the development partition had 15 sessions, and the test partition 12 sessions. There were a total of 48,000 images in total in the training partition, 45,000 in development, and 37,695 in testing (130,695 frames in total). For SEMAINE, one-minute segments of the most facially-expressive part of each selected interaction were coded. For the baseline system for this task, two types of features were extracted: two-layer appearance features (Local Binary Gabor Patterns) and geometric features derived from tracked facial point locations, which were then fed into a linear SVM. The average MSE on AU in BP4D datasets was around 0.8, while similar techniques were not applied on SEMAINE. [130, 374] took into consideration all the frames of videos, which in turn made the training more time taking.

Nicolaou et al. [248] developed an algorithm for automatically segmenting videos into data frames, in order to show the transition of emotions. To ensure oneto-one correspondence between timestamps of accorder, annotations were binned according to video frames. The crossing over from one emotional state to the other was detected by examining the valence values and identifying the points where the sign changed. The crossovers were then matched across coders. The crossover frame decision was made and the start frame of the video segment decided. The ground truth values for valence were retrieved by incrementing the initial frame number where each crossover was detected by the coders. The procedure of determining combined average values continued until the valence value crossed again to a non-negative valence value. The endpoint of the audio-visual segment was then set to the frame including the offset, after crossing back to a non-negative valence value. Discerning dimensional emotions from head gestures proposed a string-based audiovisual fusion which achieved better results for dimensions valence and expectation as compared to feature-based fusion. This approach added video-based events like facial expression action units, head nods, shakes as ‘words’ to string of acoustic events.

The non-verbal visual events were extracted similar to the unimodal analysis illustrated in [130] (use of nodeHMM and shakeHMM). For detection of facial action units, a local binary patterns descriptor was used and tested on the MMI facial Expression Database. For verbal and non-verbal acoustic events, emotionally relevant keywords derived from automatic speech recognition (ASR) transcripts of SEMAINE, were used. Key words were detected using the multi-stream large vocabulary continuous speech recognition (LVCSR) engine on recognizer's output, rather than ground truth labels. Finally, a SVR with linear kernel was trained. The event fusion was performed at the string level per segment, by joining all events where more than half of the event overlapped with the segment in a single string. The events could thus be seen as ‘words’. The resulting strings were converted to a feature vector representation through a binary bag-of-words (BOW) approach. This lead to an average correlation coefficient of 0.70 on Activation, Valence and Intensity which nearly matches human accuracy for the same task.

Recent Works on the HUMAINE Dataset

Chang et al. [56] worked on the vocal part of the HUMAINE dataset information to analyze emotion, mood and mental state, eventually combining it into low footprint C library as AMMON for phones. Sound processing starts with segmenting the audio stream from the microphone into frames with fixed duration (200 ms) and fixed stepping duration (80 ms). The features selected were LLDs (ZCR, RMS, MFCC, etc.) and functions (Mean, SD, skewness). AMMON was developed by extending an ETSI (European Tele-communications Standards Institute) front-end feature extraction library. It included features to describe glottal vibrational cycles which is a promising feature for monitoring depression. They performed a 2-way classification task to separate clips with positive emotions from those with negative emotions. A feature vector was extracted from each clip using AMMON without glottal timings. Finally, the use of SVM with these feature vectors produced 75% accuracy on BELFAST (The naturalistic dataset of HUMAINE).

Castellano et al. [53] aimed to integrate information from facial expressions, body movement, gestures and speech, for recognition of eight basic emotions. The facial features were extracted by generating feature masks, which were then used to extract feature points, comparing them to a neutral frame to produce FAPs as in the previous research. Body tracking was performed using the EyesWeb platform which tracked silhouettes and blobs, extracting motion and fluidity as main expressive cues. The speech feature extraction focused on intensity, pitch, MFCC, BSB and pause length. These were then independently fed into a Bayesian classifier and integrated at decision-level fusion. While the unimodal analysis led to an average of 55% accuracy, feature-level fusion produced a significantly higher accuracy of 78.3%. Decision-level fusion results did not vary much over feature-level fusion.

Another interesting work in [110] aims to present a novel approach to online emotion recognition from visual, speech and text data. For video labeling, temporal information was exploited, which is known to an important issue, i.e., one utterance at time (t) depends on the utterance at time t. The audio features used in the study included: signal energy, pitch, voice quality, MFCC, spectral energy and time signal, which were then modeled using a LSTM. The spoken content knowledge was incorporated at frame level via early fusion, wherein negative keywords were used for activation, and positive for valence. Subsequently, frame-based emotion recognition with unimodal and bimodal feature sets, and turn-based emotion recognition with an acoustic feature set were performed as evaluations. Finally, whilst a SVR was found to outperform a RNN in recognizing activation features, the RNN performed better in recognition of valence from frame-based models. The inclusion of linguistic features produced no monotonic trend in the system.

Recent Works on the eNTERFACE Dataset

eNTERFACE dataset is one the most widely used datasets in multimodal emotion recognition. Though in this discussion we mainly focus on multimodalities, we also explain some of the notable unimodal works which have impacted this research field radically.

Among unimodal experiments reported on this dataset, one of the notable works was carried out by Eyben et al. [111]. They pioneered the openEAR, a toolkit to extract speech related features for affect recognition. Several LLDs like Signal Energy, FFT-spectrum, MFCC, Pitch and their functionals were used as features. Multiple Data Sinks were used in the feature extractor, feeding data to different classifiers (K-Nearest Neighbor, Bayes and Support-Vector based classification and regression using the freely available LibSVM). The experiments produced a benchmark accuracy of 75% on the eNTERFACE dataset.

The study by Chetty et al. [66] aimed to develop an audiovisual fusion approach at multiple levels to resolve the misclassification of emotions that occur at unimodal level. The method was tested on two different acted corpora, DaFEx and eNTERFACE. Facial deformation features were identified using singular value decomposition (SVD) values (positive for expansion and negative for contraction) and were used to determine movement of facial regions. Marker-based audio visual features were obtained by dividing the face into several sectors, and making the nose marker the local center for each frame. PCA was used to reduce the number of features per frame to a 10-dimensional vector for each area. LDA optimized SVDF and VDF feature vectors and an SVM classifier was used for evaluating expression quantification, as High, Medium and Low. The unimodal implementation of audio features led to an overall performance accuracy of around 70% on DaFEx and 60% on eNTERFACE corpus, but the sadness-neutral pair and happiness-anger pair were confused significantly. The overall performance accuracy for visual only features was found to be around 82% for the eNTERFACE corpus and only slightly higher on the DaFEx corpus, however, a significant confusion value on neutral-happiness and sadness-anger pairs was found. Audiovisual fusion led to an improvement of 10% on both corpus, significantly decreasing the misclassification probability.

Another attempt [424] at merging audio-visual entities led to 66.5% accuracy on the eNTERFACE dataset (Anger being the highest at 81%). They adopted LBP for facial image representations for facial expression recognition. The process of LBP features extraction generally consists of three steps: first, a facial image is divided into several non-overlapping blocks. Second, LBP histograms are computed for each block. Finally, the block LBP histograms are concatenated into a single vector. As a result, the facial image is represented by the LBP code. For audio features, prosody features like pitch, intensity and quality features like HNR, jitter and MFCC are extracted. These features were fed into a SVM with the radial basis function kernel. While unimodal analysis produced an accuracy of 55% (visual at 63%), multi modal analysis increased this to 66.51%, demonstrating support for the convergence idea.

While the previous two papers focused on late fusion-based emotion recognition, SAMMI [259] was built to focus on real-time extraction, taking into account low quality videos and noise. A module called 'Dynamic Control' to adapt the various fusion algorithms and content-based concept extractors to the quality of input signals. For example, if sound quality was detected to be low, the relevance of the vocal emotional estimation, with respect to video emotional estimation, was reduced. This was an important step to make the system more reliable and lose some constraints. The visual part was tested on two approaches: (a) Facial FP absolute movements (b) Relative movements of couples of facial FP. For low cost benefits, authors used the Tomasi implementation of the Lukas Kanade (LK) algorithm (embedded in the Intel OpenCV library). The vocal expressions extracted were similar to those reported in other papers (HNR, jiter, intensity, etc.). The features were fed as one second window interval definitions, into two classifiers: SVM and a conventional Neural Network (NN). Finally, SAMMI performed fusion between estimations resulting from the different classifiers or modalities. The output of such a module significantly enhanced the system performance. Since the classification step is computationally efficient with both NN and SVM classifiers, multiple classifiers can be employed at the same time without adversely impacting the system performance. Though the NN was found to improve the CR+ value in fear and sadness, an overall Bayesian network performed equally well with a CR+ of 0.430.

Table 3.5 Comparative table: Key studies on multimodal emotion analysis datasets

Datasets	Reference	Summary	Performance
SEMAINE	Gunes et al. [130]	V	0.094 (MSE)
	Valstar et al. [374]	V	68.10% (Acc)
	Eyben et al. [113]	A + V	0.190 (MLE)
HUMAINE	Chang et al. [56]	A	93.60% (Acc)
	Castellano et al. [53]	A + V	78.30% (Acc)
	Eyben et al. [110]	A+T	0.55 (CC)
eNTERFACE	Eyben et al. [111]	A	75.20% (Acc)
	Chetty et al. [66]	T+V	86.10% (Acc)
	Zhang et al. [424]	A + V	66.51% (WAA)
	Paleari et al. [259]	A + V	43.00% (WAA)
	Dobrivsek et al. [93]	A + V	77.20% (UNWAA)
IEMOCAP	Rahaman et al. [312]	A	72.80%(A)
	Jio et al. [168]	A+T	69.20% (WAA)
	Metallinou et al. [229]	A + V	75.45%(UWAA)
	Rozgic et al. [327]	A + V	69.50%(WAA)
	Poria et al. [294]	A + T + V	76.85%(UNWAA)

Legenda: *A* Audio, *T* Text, *V* Video, *MSE* Mean Squared Error, *MLE* Maximum Likelihood Estimate, *Acc* Accuracy, *WAA* Weighted Average Accuracy, *UWAA* Unweighted Average Accuracy, *CC* Co-relation Coefficient

Recent Works on the IEMOCAP Dataset

In multimodal emotion recognition, IEMOCAP dataset is the most popular dataset and numerous works have reported its use as a benchmark. Below, we outline some of the recent key works. A summary and comparison of these studies are shown in Table 3.5.

Rehman and Busso [312] developed a personalized emotion recognition system using an unsupervised feature adaption scheme by exploiting the audio modality. The OpenSMILE toolkit with the INTERSPEECH 2009 Emotion Challenge feature set was used to extract a set of common acoustic and prosodic features. A linear kernel SVM with sequential minimal optimization (SMO) was used as the emotion detector. The purpose of normalizing acoustic features was to reduce speaker variability, while preserving the discrimination between emotional classes. The iterative feature normalization approach iteratively estimated the normalizing parameters from an unseen speaker. It thus served as a suitable framework for personalized emotion recognition system. In the IFN scheme, an emotion recognition system was used to iteratively identify neutral speech of the unseen speaker. Next, it estimated the normalization parameters using only this subset (relying on the detected labels). These normalization parameters were then applied to the entire data, including the emotional samples. To estimate the performance, the study used leave-one-speaker out, 10-fold cross validation. The results on the IEMOCAP database indicated that the accuracy of the proposed system was 2% (absolute) higher than the one achieved by the baseline, without the feature adaptation scheme. The results on uncontrolled recordings (i.e., speech downloaded from a video-sharing website) revealed that the feature adaptation scheme significantly improved the unweighted and weighted accuracies of the emotion recognition system.

While most papers have focused on audio-visual fusion, Qio Jio [168] reported emotion recognition with acoustic and lexical features. For acoustic features, low-level acoustic features were extracted at frame level on each utterance and used to generate feature representation of the entire dataset, using the OpenSMILE toolkit. The features extracted were grouped into three categories: continuous, qualitative and cepstral. Low-level feature vectors were then turned into a static feature vector. For each emotional utterance, a GMM was built via MAP adaptation using the features extracted in the same utterance. Top 600 words from each of the four emotion classes respectively were selected and merged to form a basic word vocabulary of size 2000. A new lexicon for each emotion class (in which each word has a weight indicating its inclination for expressing this emotion) was constructed. The new emotion lexicon not only collected words that appeared in one emotion class but also assigned a weight indicating its inclination for expressing this emotion. This emotion lexicon was then used to generate a vector feature representation for each utterance. Two types of fusion schemes were experimented with: early fusion (feature concatenation) and late fusion (classification score fusion). The SVM with linear kernel was used as emotion classifier. The system based on early fusion of Cepstral-BoW and GSV-mean acoustic features combined with ACO-based system, Cepstrum-based system, Lex-BoW-based system, and Lex-eVector-based system through late fusion achieves the best weighted emotion recognition accuracy of 69.2%.

Continuing with bimodal systems, Metallinou et al. [229] carried out emotion recognition using audio-visual modalities by exploiting Gaussian Mixture Models (GMMs). Markers were placed on the faces of actors to collect spatial information of these markers for each video frame in IEMOCAP. Facial markers were separated into six blocks, each of which defined a different facial region. A GMM was trained for each of the emotional states examined; angry (ANG), happy (HAP), neutral (NEU) and sad (SAD). The marker point coordinates were used as features for the training of Gaussian mixture models. The frame rate of the markers was 8.3 ms. The feature vector for each facial region consisted of 3-D coordinates of the markers belonging to that region plus their first and second derivatives. GMM with 64 mixtures was chosen as it was shown to achieve good performance. MFCCs are used for vocal analysis. The feature vector comprised 12 MFCCs and energy, their first and second derivatives; constituting a 39-dimensional feature vector. The window length for the MFCC extraction was 50 ms and the overlap set to 25 ms, to match the window of the facial data extraction. Similar to facial analysis, a GMM was trained for each emotion along with an extra one for background noise. Here, a GMM with 32 mixtures was chosen. Two different classifier combination techniques were explored: the first a Bayesian approach for multiple cue combination, and the second an ad-hoc method utilizing SVMs with radial basis kernels that used post classification accuracies as features. Anger and happiness were found to have better recognition accuracies in the face-based classifier compared to emotional states with lower levels of activation, such as sadness and neutrality; while anger and sadness demonstrated good accuracy in voice-based classifiers. A support vector classifier (SVC) was used to combine the separate face and voice model decisions. The Bayesian classifier and SVC classifiers were found to perform comparably, with neutral being the worst recognized emotional state, and anger/sadness being the best.

While previous works focused on bimodality, the work in [327] aims to classify emotions using audio, visual and textual information by attaching probabilities to each category based on automatically generated trees, with SVMs acting as nodes. There were several acoustic features used, ranging from jitter and shimmer for negative emotions to intensity and voicing statistics per frame. Instead of representing the non-stationary MFCC features using statistical functionals as in previous works, they use a set of model-based features obtained by scoring all MFCC vectors in a sentence using emotion-dependent Gaussian mixture models (GMM). The lexical features were summarized using LIWC and GI systems represented by bag of word stems. The visual features encapsulated facial animation parameters representing nose, mouth and chin markers, eyebrow angle, etc. A randomized tree is generated using the set of all classifiers whose performance is above a threshold parameter. The experiments were conducted in leave-one-speaker-out fashion. The unimodal feature set achieved an accuracy of around 63% whereas their combination led to an increase of around 8%.

3.4.2.3 Other Multimodal Cognitive Research

DeVault et al. [87] introduced SimSensei Kiosk, a virtual human interviewer named Ellie, for automatic assessment of distress indicators among humans. Distress indicators are verbal and non-verbal behaviors correlated with depression, anxiety or post-traumatic stress disorder (PSTD). The SimSensei Kiosk was developed in such a way the user feels comfortable talking and sharing information, thus providing clinicians an automatic assessment of psychological distress in a person. The evaluation of the kiosk was carried out by the Wizard-of-Oz prototype system, which had two human operators for deciding verbal and non-verbal responses. This development of SimSensei kiosk was carried out over a period of two years with 351 participants, out of which 217 were male, 132 were female and 2 did not report the gender. In this work, termed the Multisense framework, a multimodal real-time sensing system was used, for synchronized capture of different modalities, real-time tracking and fusion process. The multimodal system was also integrated with GAVAM head tracker, CLM-Z face tracker, SHORE face detector, and more.The SimSensei Kiosk used 4 statistically trained utterance classifiers to capture the utterance meaning of the users and Cerebella, a research platform for realization of the relation between mental states and human behavior.

Alam et al. [2], proposed an automatic personality trait recognition framework using the YouTube personality dataset. The dataset consists of videos by 404 YouTube bloggers (194 male and 204 female). The features used for this task were linguistic, psycholinguistic, emotional features and audio-visual features. Automatic recognition of personality traits is an important topic in the field of NLP, particularly aimed at processing the interaction between human and virtual agents. High dimensional features were selected using the relief algorithm and classification models were generated using SMO for the SVM. At the final stage, decision-level fusion for classification of personality traits was used.

Other notable work in personality recognition was carried out by Sarkar et al. [332], who used the YouTube personality dataset and a logistic regression model with ridge estimator, for classification purposes. They divided features into five categories, i.e., audio-visual features, text features, word statistics features, sentiment features and gender features. A total of 1079 features were used, with 25 audiovisual features, 3 word statistics feature, 5 sentiment feature, 1 demographic feature and 1045 text features. In conclusion, their in-depth feature analysis showcased helpful insights for solving the multimodal personality recognition task.

Siddiquie et al. [356], introduced the task of exploiting multimodal affect and semantics for automatic classification of politically persuasive web videos. Rallying A Crowd (RAC) dataset was used for experimentation with 230 videos. The approach was executed by extraction of audio, visual and textual features to capture affect and semantics in the audio-video content and sentiment in the viewers' comments. For the audio domain, several grades of speech arousal and related semantic categories such as crowd reaction and music were detected. For the visual domain, visual sentiment and semantic content were detected. The research employed both feature-level and decision-level fusion methods. In the case of decision-level fusion, the author used both conventional- and learning-based decision fusion approaches to enhance the overall classification performance.

In Table 3.6, some key research works in multimodal sentiment analysis and opinion mining are summarized and categorized based on their proposed method.

3.5 Available APIs

In this section, we list 20 popular APIs for emotion recognition from photos, videos, text and speech. The main categories of emotions that are detected using the APIs are Joy, Anger, Contempt, Fear, Surprise, Sadness and Disgust.

Sentiment analysis is also explored by some of the APIs, in addition to emotion recognition, to determine whether the expressed emotion is positive or negative.

Table 3.6 Study characteristics of recent papers on multimodal analysis

References	Modality	Speakers/Datasets	Model	Features	Fusion type
DeVault et al. [87]	A + T + V	351 (217 M, 132 F and 2 did not not report the gender)	4 statistically trained utterance classifiers	Smile intensity, 3D head position and orientation, intensity or lack of facial expressions like anger, disgust and joy, speaking fraction, dynamics, speech dynamics, gaze direction, etc.	MF
Alam et al. [2]	A + T + V	404 YouTube vloggers (194 M, 210 F)/ YouTube personality dataset	SMO for SVM	A-V, lexical, POS psycholinguistic, emotional and traits	D
Sarkar et al. [332]	A + T + V	404 YouTube vloggers/ YouTube personality dataset	LR model with ridge estimator	A-V, text, demographic and sentiment	N/A
Siddiquie et al. [356]	A + T + V	230 videos/Rallying a crowd (RAC) dataset	RBF SVM and LR classifier	Softwares using CAFFEE and features (prosody, MFCC, or spectrogram) and using SATSVM and DCM	F and D

Legenda: *A* Audio, *V* Video, *T* Text, *ML* Machine Learning, *SMO* Sequential Minimal Optimization, *LR* Logistic Regression, *MKL* Multiple Kernel Learning, *D* Decision, *F* Feature, *MF* Multisense Framework

- *Emotient*[1] detects Attention, Engagement and Sentiment from facial expressions. These factors are considered key performance indicators for adding business value to advertising, media, consumer packaged goods and other industries, which need consumers' feedbacks to improve the quality of their products.
- *Imotions*[2] combines Emotient face expression technology to extract emotions and sentiments from various observed bodily cues. It can also be easily combined with other technologies such as EEG, eye tracking, Galvanic Skin response, etc., to improve emotion recognition accuracy.
- *EmoVu*[3] by Eyeris is a comprehensive face analytics API that employs deep learning for emotion recognition. The API also provides vision software to support ambient intelligence and is also useful for detecting age and gender identification, eye tracking and gaze estimation.
- *nViso*[4] uses 3D facial imaging technology for emotion recognition from facial expressions in real time. The software is completely automated and received the IBM award for smarter computing in 2013.
- *Alchemy API*[5] is also powered by IBM Watson. The API performs sentiment analysis on large and small documents, news articles, blog posts, product reviews, comments and tweets.
- *Kairos*[6] provides an API for analyzing facial expressions and features for emotion recognition, gender and age detection and attention management. It provides applications for various industries such as advertising, market research, health, financial services, retail, etc.
- *Tone API*[7] provides emotional insights from written text. It focuses mainly on marketers and writers to improve their content on the basis of emotional insights.
- *Project Oxford*[8] by Microsoft provides APIs for categories such as Vision, Speech, Language, Knowledge and Search.
- *Face reader*[9] by Noldus is widely used for academic purposes. It is a facial expression analysis software for analyzing universal emotions in addition to neutral and contempt. The software is also used to observe gaze direction and head orientation.
- *Sightcorp*[10] is a facial expression analysis API and is also used for eye tracking, age and gender estimation, head pose estimation, etc.

[1] http://emotient.com

[2] http://imotions.com

[3] http://emovu.com

[4] http://nviso.ch

[5] http://alchemyapi.com

[6] http://kairos.com

[7] http://toneapi.com

[8] http://microsoft.com/cognitive-services/en-us/apis

[9] http://noldus.com/facereader

[10] http://sightcorp.com

- *SkyBiometry*[11] is a face recognition and face detection, cloud biometrics API. This API is used to detect emotions such as happy, sad, angry, surprise, disgust, scared and neutral from faces.
- *CrowdEmotions*[12] detects the dynamics of six basic emotions: happiness, surprise, anger, disgust, fear and sadness. It also captures people's engagement, emotions and body language, towards a particular event.
- *Affectiva*[13] is an API for emotion recognition using deep learning. It is said to have nearly 4 million faces as emotion database in order to provide great accuracy.
- *The Tone Analyzer*[14] is an API, powered by IBM Watson, for analyzing emotional states in text.
- *Repustate API*[15] is used for sentiment analysis in text. This API is based on linguistic theory and review cues based on POS tagging, lemmatization, prior polarity and negations.
- *Receptiviti API*[16] is used to analyze texts, tweets, emails, chats, surveys and voice data to provide insights into various aspects of people's personal lives, such as personality, emotion, tone and relationships.
- *Bitext*[17] is a text analysis API that is used for sentiment analysis, categorization, entity extraction and concept extraction. It is mainly focused for market research specialists.
- *Mood Patrol*[18] is used to detect emotions from given text. It was developed by Soul Hackers Lab and works reasonably well on small documents.
- *Synesketch*[19] is an open source software used for textual emotion recognition, sentiment recognition and visualization. It analyzes text in terms of emotions such as happiness, sadness, anger, fear, disgust, and surprise, and the intensity of emotion and sentiment, such as positive or negative.
- *Sentic API*[20] is a free API for emotion recognition and sentiment analysis providing semantics and sentics associated with 50,000 commonsense concepts in 40 different languages.

[11] http://skybiometry.com

[12] http://crowdemotion.co.uk

[13] http://affectiva.com

[14] http://tone-analyzer-demo.mybluemix.net

[15] http://repustate.com/sentiment-analysis

[16] http://receptiviti.ai

[17] http://bitext.com/text-analysis-api

[18] http://market.mashape.com/soulhackerslabs/moodpatrol

[19] http://krcadinac.com/synesketch

[20] http://sentic.net/api

3.6 Discussion

Timely surveys are rudimentary for any field of research. In this survey, we not only discuss the state of the art but also collate available datasets and illustrate key steps involved in a multimodal affect analysis framework. We have covered around 100 papers in our study. In this section, we describe some of our major findings from this survey.

3.6.1 Major Findings

The multimodal analysis of affective content now a days is as popular as the unimodal analysis. This is due to the need of mining information from the growing amount of videos posted the social media and the advancement of human-computer interaction agents. As discussed in [92], the trends in multimodal affect analysis can be classified into two timelines. Till 2003, "the use of basic signal processing and machine learning techniques, independently applied to still frames (but occasionally to sequences) of facial or vocal data, to detect exaggerated context free expressions of a few basic affective states, that are acted by a small number of individuals with no emphasis on generalizability" – in other words, mainly unimodal and in some cases bimodal, i.e., audio-visual clues, were used for affect analysis. In 2016, the trend leans towards using more than one modality for affect recognition of videos using machine learning techniques. In particular, there has been a growing interest in using deep learning techniques and a number of fusion methods. There is a significant amount of work that has been done in multimodal sentiment analysis in the past 3 years. The types of dataset are radically changing, wherein the past, acted data were being used, but presently, videos are being crawled from YouTube and used for experimentation in research.

Though visual and audio modalities have been used for multimodal affect recognition in many studies since 2004 (Table 3.4), it is worth mentioning that although most of the reported works use audio and visual information for affect recognition, recent advancements in text affect analysis [292] have led to increasing use of text modality in these works, particularly from 2010 onwards [239]. For example, from the Table 3.3 it can be seen that from 2010 onwards, text modality has been considered in many research works on multimodal affect analysis.

In our literature survey, we have found more than 90% of studies reported visual modality as superior to audio and other modalities. Audio modality often suffers from the presence of noise in the signal. However, recent studies on multimodal sentiment analysis by Perez et al. [273] have demonstrated that an efficient and intelligent text analysis engine can outperform other unimodal classifiers, i.e., visual and audio. In both these independent studies, text modality was found to play the most vital role in multimodal sentiment analysis, and, furthermore, when fused with audio-visual features, it was shown to improve the performance significantly. On

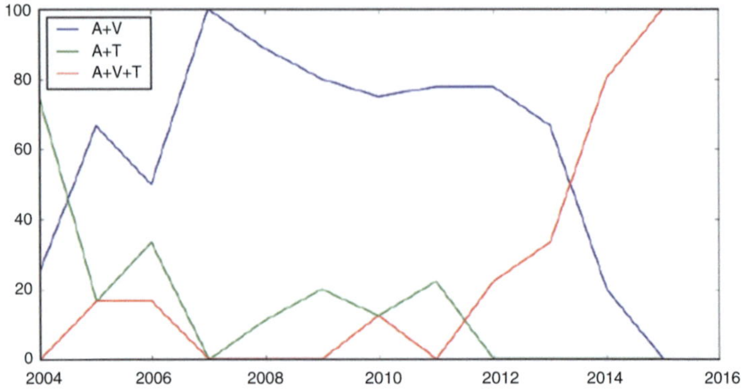

Fig. 3.5 Percentage of research works done using different modalities for affect Recognition over the years (Legenda: *A* Audio, *T* Text, *V* Video)

the MOUD dataset Perez et al. [273] obtained accuracies of 70.94%, 67.31% and 64.85% respectively for textual, visual and audio modalities (Fig. 3.5).

In summary, as noted earlier, there are several concerns that need to be addressed in this research field. Firstly, the amount of trust that should be placed in acted corpora is debatable. The primary question that arises is if they appropriately replicate the natural characteristics of a spontaneous expression. For example, in acted data, people rarely smile while acting as a frustrated person, whereas studies [148] show that in 90% of cases in real-life situations, people smile while expressing their frustration. Such errors eventually lead to poor generalization of any multimodal sentiment analysis system.

Apart from replication, the taxonomy of the emotions and sentiments is never set in stone. Though for sentiment analysis, it is relatively straight forward and practically convenient to use positive, negative and neutral sentiment dimensions, in the case of emotions, the number of emotional dimensions to use is unclear. In the literature, most studies use Ekman's six basic emotions, i.e., Anger, Disgust, Joy, Surprise, fear and Sad, for experimentation, however people often tend to use complex emotions like love, frustration, etc., in their day to day conversations.

For text modality, deep learning is now a days the most popular methods. Compare to the commonly used bag of words method bag of concepts based methods are also developed by researchers [384]. On the other hand, also for visual modality the trend has shifted from the use of different complex image processing methods to the development of complex deep networks. With the advent of CNN [406], C3D [371] the video classification performance using the deep networks has overshadowed existing image processing algorithms like Optical flow, ASM, AAM. Though deep learning based methods are quite popular in text and visual affect recognition, not many works have been proposed in the literature for audio classification using deep networks. So, for audio classification the hand-crafted feature computation methods, e.g., OpenSMILE [337], are still very popular and widely used in the audio affect classification research. The degree to which

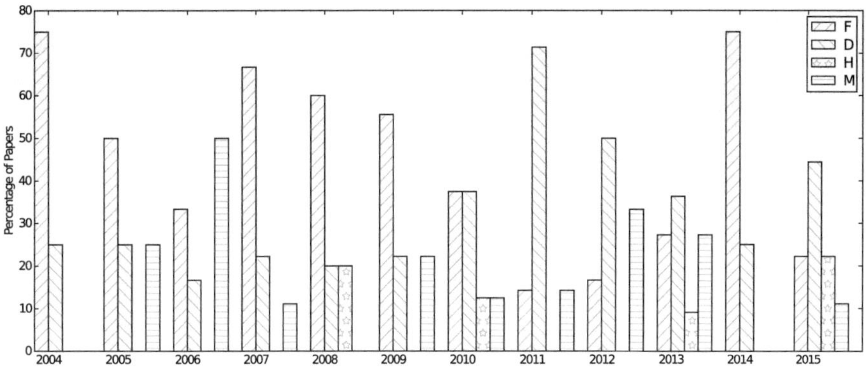

Fig. 3.6 Percentage of research articles in multimodal affect analysis using different fusion methods over the years (Legenda: F = Feature-Level Fusion; D = Decision-Level Fusion; H = Hybrid Fusion; M = Model-Based Fusion)

a multimodal system can be generalized is also a crucial factor in determining its practical implementation. For example, it is particularly difficult to determine whether a developed approach is subject independent and can work well with any context, and to what extent the system should be trained on diverse contextual data.

To date, the most widely used fusion method is feature-level fusion, which consumes a lot of time and requires effective feature selection methods. Since 2010, multimodal fusion has drawn increasing attention of researchers, and a number of decision-level fusion methods have been recently reported, as can be seen from Fig. 3.6. Multimodal fusion is useful not only in sentiment analysis but it has shown its usefulness in a wide area of multimedia analytics problems [341, 342] such as content uploading [137, 343], lecture video segmentation [346, 347], soundtrack recommendation for outdoor videos [350, 351], event detection and summarization [345, 349], tag recommendation and ranking [282, 295, 304, 344, 348], SMS and MMS based FAQ retrieval [352, 353], map matching [411], and precision agriculture [175].

3.7 Conclusion

In this chapter, we carried out a review of the fundamental stages of a multimodal affect recognition framework. We started by discussing available benchmark datasets, followed by an overview of the state of the art in audio-, visual- and textual-based affect recognition. In particular, we highlighted prominent studies in unimodal affect recognition, which we consider crucial components of a multimodal affect detector framework. For example, without efficient unimodal affect classifiers or feature extractors, it is not possible to build a well-performing multimodal affect detector. Hence, if one is aware of the state of the art in unimodal affect recognition, which has been thoroughly reviewed in this chapter, it would facilitate the construction of an appropriate multimodal framework. Our survey has confirmed other researchers' findings that multimodal classifiers can outperform unimodal classifiers. Furthermore, text modality plays an important role in boosting the performance of an audio-visual affect detector. On the other hand, the use of deep learning is increasing in popularity, particularly for extracting features from modalities. Although feature-level fusion is widely used for multimodal fusion, there are other fusion methods developed in the literature.

In the following chapters, we present text and multimodal sentiment analysis methods built upon this literature. The following chapters also address number of drawbacks in the existing methodologies and propose algorithms to overcome them.

Chapter 4
Concept Extraction from Natural Text for Concept Level Text Analysis

Abstract Concept extraction from text is a key step in concept-level text analysis. In this chapter, we propose a ConceptNet-based semantic parser that deconstructs natural language text into concepts based on the dependency relation between clauses. Our approach is domain-independent and is able to extract concepts from heterogeneous text. Through this parsing technique, 92.21was obtained on a dataset of 3,204 concepts. We also show experimental results on three different text analysis tasks, on which the proposed framework outperformed state-of-the-art parsing techniques.

Keywords Emotion recognition · Sentiment analysis · Dependency relation · Natural language processing

4.1 Introduction

Concept-level text analysis [300, 303] focuses on a semantic analysis of text through the use of web ontologies or semantic networks, which allow the aggregation of conceptual and affective information associated with natural language opinions. By relying on large semantic knowledge bases, such approaches step away from blind use of keywords and word co-occurrence count, but rather rely on the implicit features associated with natural language concepts. Unlike purely syntactical techniques, concept-based approaches are able to detect also sentiments that are expressed in a subtle manner, e.g., through the analysis of concepts that do not explicitly convey any emotion, but which are implicitly linked to other concepts that do so. The bag-of-concepts model can represent semantics associated with natural language much better than bags-of-words. In the bag-of-words model, in fact, a concept such as `cloud computing` would be split into two separate words, disrupting the semantics of the input sentence (in which, for example, the word `cloud` could wrongly activate concepts related to `weather`).

The analysis at concept-level allows for the inference of semantic and affective information associated with natural language opinions and, hence, enables a comparative fine-grained feature-based sentiment analysis. Rather than gathering

S. Poria et al., *Multimodal Sentiment Analysis*, Socio-Affective Computing 8,
https://doi.org/10.1007/978-3-319-95020-4_4

isolated opinions about a whole item (e.g., iPhone5), users are generally more interested in comparing different products according to their specific features (e.g., iPhone5's vs Galaxy S3's touchscreen), or even sub-features (e.g., fragility of iPhone5's vs Galaxy S3's touchscreen). In this context, the construction of comprehensive common and common-sense knowledge bases is key for feature-spotting and polarity detection, respectively. Common-sense, in particular, is necessary to properly deconstruct natural language text into sentiments; for example, to appraise the concept `small room` as negative for a hotel review and `small queue` as positive for a post office, or the concept `go read the book` as positive for a book review but negative for a movie review [39]. Common-sense knowledge describes basic understandings that people acquire through experience. In cognitive science, building conceptual representations is a fundamental ability to understand and handle objects and actors of an operating environment.

To this end, the propose concept parser aims to break text into clauses and, hence, deconstruct such clauses into concepts, to be later fed to a vector space of common-sense knowledge. For applications in fields such as real-time human-computer interaction and big social data analysis, in fact, deep natural language understanding is not strictly required: a sense of the semantics associated with text and some extra information (e.g., affect) associated with such semantics are often enough to quickly perform tasks such as emotion recognition and polarity detection. Common-sense reasoning is often performed through common-sense ontologies and the employment of reasoning algorithms, such as predicate logic and machine learning, to reach a conclusion.

4.2 The Patterns for Concept Extraction

Below the patterns for concept extraction are explained. The patterns are built based on the dependency tree of a sentence.

4.2.1 Subject Noun Rule

Trigger: when the active token is found to be the syntactic subject of a verb.
Behavior: if a word *h* is in a subject noun relationship with a word *t* then the concept *t-h* is extracted.
Example: In (1), *movie* is in a subject relation with *boring*.

(1) The movie is boring.

Here the concept `(boring-movie)` is extracted.

4.2.2 Joint Subject Noun and Adjective Complement Rule

Trigger: when the active token is found to be the syntactic subject of a verb and the verb is on adjective complement relation with an adverb.
Behavior: if a word *h* is in a subject noun relationship with a word *t* and the word *t* is with adjective complement relationship with a word *w* then the concept *w-h* is extracted.
Example: In (2), *flower* is in a subject relation with *smells* and *smells* is in adjective complement relationship with *bad*.

(2) The flower smells bad.

Here the concept `(bad-flower)` is extracted.

4.2.3 Direct Nominal Objects

This complex rule dealt with direct nominal objects of a verb.

Trigger: when the active token is head verb of a direct object dependency relation.
Behavior: if a word *h* is in a direct nominal object relationship with a word *t* then the concept *h-t* is extracted.
Example: In (3) the system extracted the concept `(see,movie)`.

(3) Paul saw the movie in 3D.

`(see,in,3D)` is not treated at this stage since it will later be treated by the standard rule for prepositional attachment.

4.2.4 Adjective and Clausal Complements Rules

These rules dealt with verbs which had as complements either an adjective or a closed clause (i.e. a clause, usually finite, with its own subject).

Trigger: when the active token is head verb of one of the complement relations.
Behavior: if a word *h* is in a direct nominal object relationship with a word *t* then the concept *h-t* is extracted.
Example: in (4), *smells* is the head of a clausal complement dependency relation with *bad* as the dependent.

(4) This meal smells bad.

In this example the concept `(smell,bad)` is extracted.

4.2.5 *Negation*

Negation is also a crucial component of natural language text which usually flips the meaning of the text. This rule is used to identify whether a word is negated in the text.

Trigger: when in a text a word is negated.
Behavior: if a word *h* is negation by a *negation marker t* then the concept *t-h* is extracted.
Example: in (5), *like* is the head of the negation dependency relation with *not* as the dependent. Here, *like* is negated by the negation marker *not*.

(5) I do not like the movie.

Based on the rule described above the concept `(not, like)` is extracted.

4.2.6 *Open Clausal Complements*

Open clausal complements are clausal complements of a verb that do not have their own subject, meaning that they (usually) share their subjects with that of the matrix clause. The corresponding rule is complex in the same way as the one for direct objects.

Trigger: when the active token is the head of the relation
Behavior: as for the case of direct objects, the algorithm tried to determine the structure of the dependent of the head verb. Here the dependent is itself a verb, therefore, the system tried to establish whether the dependent verb had a direct object or a clausal complement of its own. In a nutshell, the system is dealing with three elements: the head verb(h), the dependent verb(d), and the (optional) complement of the dependent verb (t). Once these elements had all been identified, the concept (h,d,t) is extracted
Example: in (6), *like* is the head of the *open clausal complements* dependency relation with *praise* as the dependent and the complement of the dependent verb *praise* is *movie*.

(6) Paul likes to praise good movies.

So, in this example the concept `(like,praise,movie)` is extracted.

4.2.7 *Adjectival, Adverbial and Participial Modification*

The rules for items modified by adjectives, adverbs or participles all share the same format.

Trigger: these rules were activated when the active token is modified by an adjective, an adverb or a participle.
Behavior: if a word *w* is modified by a word *t* then the concept *(t,w)* is extracted.
Example: in (7) the concept *bad, loser* is extracted.

(7) a. Paul is a bad loser.

4.2.8 *Prepositional Phrases*

Although prepositional phrases do not always act as modifiers, we introduced them in this section as the distinction didn't really matter for their treatment.

Trigger: the rule is activated when the active token is recognized as typing a prepositional dependency relation. In this case, the head of the relation is the element to which the PP attached, and the dependent is the head of the phrase embedded in the PP.
Behavior: instead of looking for the complex concept formed by the head and dependent of the relation, the system used the preposition to build a ternary concept.
Example: in (8), the parser yields a dependency relation typed `prep_with` between the verb *hit* and the noun *hammer* (=the head of the phrase embedded in the PP).

(8) Bob hit Marie with a hammer.

Therefore, the system extracted the complex concept `(hit, with, hammer)`.

4.2.9 *Adverbial Clause Modifier*

This kind of dependency concerned full clauses that act as modifiers of a verb. Standard examples involved temporal clauses and conditional structures.

Trigger: the rule is activated when the active token is a verb modified by an adverbial clause. The dependent is the head of the modifying clause.
Behavior: if a word *t* is a adverbial clause modifier of a word *w* then the concept *(t-w)* is extracted.
Example: in (9), the complex concept `(play,slow)` is extracted.

(9) The machine slows down when the best games are playing.

4.2.10 Noun Compound Modifier

Trigger: the rule is activated when it finds a noun composed with several nouns. A noun compound modifier of an NP is any noun that served to modify the head noun.

Behavior: if a noun-word *w* is modified by another noun-word *t* then the complex concept *(t-h)* is extracted.

Example: in (10), the complex concept `(birthday,party)` is extracted.

(10) Erik threw the birthday party for his girlfriend.

4.2.11 Single Word Concepts

Words having part-of-speech Verb, Noun, Adjective and Adverb were also extracted from the text. Single word concepts which existed in the multi-word-concepts are discarded as they carry redundant information. For example, concept *party* that already appeared in the concept *birthday party* so we discarded the concept *party*.

4.3 Experiments and Results

To calculate the performance, we selected 300 sentences from the *Stanford Sentiment Dataset* [360] and extracted the concepts manually. This process yielded 3204 concepts. Below in Table 4.1 we show the accuracy of concept mining process using approach with the POS based approach [313].

4.4 Conclusion

The novel dependency tree based linguistic patterns are proposed in this chapter to extract the concepts. These concepts are the fundamental units for concept level text analysis [42]. The evaluation shows the proposed method performs better than the state of the art [313]. However, this chapter does not focus on showing the performance concept extraction process in different concept level text analysis tasks. The following chapters will focus on that aspect by showing the performance of the concept level emotion recognition and sentiment analysis methods.

Table 4.1 Results obtained using different algorithms on the dataset

Algorithm	Precision
Part-of-speech approach	86.10%
Proposed approach	92.21%

Chapter 5
EmoSenticSpace: Dense Concept-Based Affective Features with Common-Sense Knowledge

Abstract This chapter proposes EmoSenticSpace, a new framework for affective common-sense reasoning that extends WordNet-Affect and SenticNet by providing both emotion labels and polarity scores for a large set of natural language concepts. The framework is built by means of fuzzy c-means clustering and support-vector-machine classification, and takes into account a number of similarity measures, including point-wise mutual information and emotional affinity. EmoSenticSpace was tested on three emotion-related natural language processing tasks, namely sentiment analysis, emotion recognition, and personality detection. In all cases, the proposed framework outperforms the state-of-the-art. In particular, the direct evaluation of EmoSenticSpace against psychological features provided in the benchmark ISEAR dataset shows a 92.15

Keywords Sentic computing · Opinion mining · Sentiment analysis · Emotion recognition · Personality detection · Fuzzy clustering

5.1 Introduction

The basic "feeling" about something can be described on a scale of approval or disapproval, good or bad, positive or negative, termed polarity. The basic question of whether customers or voters are satisfied with a product, service or action can be answered by detecting the average polarity of what they express in blogs and reviews. For such a purpose, many online resources have been developed in recent years. SenticNet, in particular, is a semantic resource for concept-level sentiment analysis built by means of dimensionality reduction [45]. It provides polarity scores for 5,700 multi-word concepts, and in its most recent release the set is expanded to 30,000 concepts. The score ranges between −1 (bad) to +1 (good),

Part of this chapter is reprinted from Knowledge-Based Systems, 69, 108-123, Poria, Gelbukh, Cambria, Hussain, Huang, "EmoSenticSpace: A novel framework for affective commonsense reasoning" 2014, with permission from Elsevier.

S. Poria et al., *Multimodal Sentiment Analysis*, Socio-Affective Computing 8,
https://doi.org/10.1007/978-3-319-95020-4_5

with neutral scores being around zero; e.g., aggravation: −0.925, accomplish goal: +0.967, December: +0.111. For example, this dictionary allows us to detect whether a customer review is 0.567 positive or 0.876 negative. However, more detailed information is often desirable [279]. Do citizens disliking the governing party specifically feel anger, fear, disgust, or shame? Do customers like a product due to the joy a product gives them or because it surprised them, and if both, are they more surprised than joyful, or vice versa. One of the main lexical resources employed to detect emotions in text is WordNet-Affect (WNA) [366]. For a relatively small set of words, WNA indicates whether a given term is related to one of six basic emotions: anger, fear, disgust, sadness, surprise, or joy. However, it does not provide information on how strong the emotion is: e.g., angered and infuriated appear as synonymous in WNA, while they evidently express a different emotional intensity. As shown later in the chapter, augmenting emotion labels with quantitative scores that go beyond the current yes/no labeling is very useful to accurately and quantitatively answer many opinion-related questions. On the other hand, for each concept SenticNet gives a quantitative measure of some unspecified emotion or unspecified mixture of emotions. However, for better opinion mining and sentiment analysis, both a quantitative measure and a specific emotion label are needed for each concept. This chapter starts with a description of an empirical method for automatically merging WNA and SenticNet, thus creating a resource, termed EmoSenticNet (ESN), with both qualitative emotion labels and quantitative polarity scores assigned to a large number of concepts [303]. Then, we present EmoSenticSpace, obtained by blending ESN with an existing resource, ConceptNet.

First (Sects. 5.3, 5.4, 5.5, and 5.6) we present and evaluate an empirical method for automatically assigning emotion labels to each concept of SenticNet, in addition to the polarity score already present in SenticNet. Associating affective information with concepts improves opinion mining results and enables mining for more detailed affective information. Then we present the EmoSenticSpace resource and show how our new resources, in combination with existing resources, offer superior accuracy on sentiment analysis, emotion detection, and personality detection tasks. Specifically, for the first task – assigning emotion labels to concepts of SenticNet, we extended emotion labels from the seed concepts (for which associated labels were known from the WNA), to all concepts in SenticNet. Thus, our task was to classify the concepts present in SenticNet into six categories given in WNA: anger, fear, disgust, sadness, surprise, and joy [298]. For this, we used a supervised learning approach. As a source of features, we used various lexical resources, most importantly the International Survey of Emotion Antecedents and Reactions (ISEAR) dataset [336], as well as WordNet. We constructed two kinds of feature: features directly assigned to a single concept and similarity measures between concepts. Each similarity measure was converted into a set of features: for a given concept, we considered its similarity to every other concept as its independent features [302]. A concept, e.g., succumb, can trigger a mix of multiple emotions, such as shame, fear, and sadness. This suggests that the problem of concept emotion identification is intrinsically a multi-label classification problem. We accounted for this intuitively by using fuzzy c-means clustering [22] to assign each concept a vector of its membership values in each of the six categories. However, our goal was single-label classification. This was partly justified by the simplicity of use of the created resource. More importantly, the obtained multi-label classification was still not reliable enough to be used in applications. Thus, we did not directly use the obtained membership values. Instead, we used them to improve the accuracy of the conventional, hard classification problem. A trivial approach would be to choose the label with the maximum membership function as the final label for a concept: succumb invokes more shame than sadness or fear. However, since the obtained membership functions were not very accurate, such a choice would not be reliable when a concept was associated with more than one category in comparable degree. Instead, we restricted the choice to the best and the second-best prediction of the fuzzy clustering (we also tried K best options, and K = 2 gave the best results). To disambiguate between these options, we used the well-known support vector machine (SVM) classifier [169]. It operated on the same set of features as the fuzzy clustering algorithm, plus the obtained fuzzy membership values as additional features. These additional features supposedly made the choice easy when the difference between the best and the second-best membership values was big enough, and when it was not, the algorithm resorted to other features to make the choice. In addition, we improved the objective function of fuzzy c-means clustering by incorporating additional functions such as point-wise mutual information (PMI) and the emotional affinity between two concepts.

5.1.1 The Algorithm for Assigning Emotion Labels to SenticNet Concepts

In summary, our algorithm for assigning emotion labels to SenticNet concepts can be outlined as below:

- Select a repertoire of concepts (as described in Sect. 5.2.1).
- Extract concepts from text (Chap. 4).
- Compute features for each concept (Sect. 5.3).
- Use these features to cluster the concepts into six fuzzy clusters (Sect. 5.8.2).
- Associate these clusters with specific emotion labels (Sect. 5.8).
- For each concept, use the fuzzy clustering results to:

  1. restrict the confusion set to top two labels;
  2. extend the feature vector by the six membership values.

- Use a hard classifier (SVM) on these extended feature vectors to disambiguate between these two labels.

5.1.2 Role of Fuzzy Clustering in Assigning Emotion Labels

For each concept, fuzzy clustering provides a vector of the membership values of a concept, i.e., affinity of a concept to each of the six emotion clusters. This is similar to the Hourglass model introduced by [42]. While the fuzzy clustering algorithm produces fuzzy clusters containing the concepts, our purpose was to identify the definite emotion class labels for each concept. The concepts that belong to more than one cluster to a significant degree, require an effective separation algorithm in order to be classified accurately. We employed the SVM-based classifier to identify the final class for each concept. Hwanjo Yu et al. [415], Mamoun Awad et al. [11], and Daniel Boley et al. [27], amongst others, have shown that clustering techniques can help to decrease the complexity of SVM training. However, these techniques consume significant computational resources to build the hierarchical structure. Jair Cervantes et al. [54] introduced SVM classification based on fuzzy clustering. In this work we follow a similar approach for emotion classification. As outlined in Sect. 5.1.1, the fuzzy clustering helps the final classification task in two ways: First, it reduces the confusion set for the SVM-based classification from 6 to 2 labels associated with the highest membership values, given that we identify the emotion labels of the corresponding clusters. Thus we reduce the task to binary classification, at which SVM is particularly effective. Second, we employ the vector of the fuzzy membership values of a particular concept as an additional feature for SVM. For example, if the fuzzy membership vector for a concept is 0.45 for anger, 0.34 for sadness, 0.03 for surprise, 0.05 for joy, 0.01 for disgust, and 0.127 for fear, then these six numeric values are used as independent features in the feature vector for this concept, along with all other features. Specific class names were obtained by employing the majority voting method described in Sect. 5.4.3. As demonstrated in Sect. 5.8.1, the reduction of the confusion set and the additional fuzzy vector features increased the classification accuracy.

5.2 Lexical Resources Used

In this section, we describe the lexical resources used to build the set of concepts, construct the features of the concepts along with the similarity measures, and to evaluate the obtained resource.

5.2.1 SenticNet

As the target lexicon and the source of polarity information for our polarity-based concept similarity measure, we used SenticNet, a freely available knowledge base that assigns polarity values to words or multi-word concepts. Specifically, we

```xml
<?xml version="1.0" encoding="UTF-8"?>
<rdf:RDF xmlns:rdf="http://www.w3.org/1999/02/22-rdf-syntax-ns#">
    <rdf:Description rdf:about="http://sentic.net/api/en/concept/celebrate_special_occasion">
        <rdf:type rdf:resource="http://sentic.net/api/concept"/>
        <text xmlns="http://sentic.net/api">celebrate special occasion</text>
        <semantics xmlns="http://sentic.net/api" rdf:resource="http://sentic.net/api/en/concept/celebrate_holiday"/>
        <semantics xmlns="http://sentic.net/api" rdf:resource="http://sentic.net/api/en/concept/celebrate_occasion"/>
        <semantics xmlns="http://sentic.net/api" rdf:resource="http://sentic.net/api/en/concept/celebrate_birthday"/>
        <semantics xmlns="http://sentic.net/api" rdf:resource="http://sentic.net/api/en/concept/celebrate_wed"/>
        <semantics xmlns="http://sentic.net/api" rdf:resource="http://sentic.net/api/en/concept/express_appreciation"/>
        <pleasantness xmlns="http://sentic.net/api" rdf:datatype="http://www.w3.org/2001/XMLSchema#float">0.93</pleasantness>
        <attention xmlns="http://sentic.net/api" rdf:datatype="http://www.w3.org/2001/XMLSchema#float">0.724</attention>
        <sensitivity xmlns="http://sentic.net/api" rdf:datatype="http://www.w3.org/2001/XMLSchema#float">0</sensitivity>
        <aptitude xmlns="http://sentic.net/api" rdf:datatype="http://www.w3.org/2001/XMLSchema#float">0</aptitude>
        <polarity xmlns="http://sentic.net/api" rdf:datatype="http://www.w3.org/2001/XMLSchema#float">0.551</polarity>
    </rdf:Description>
</rdf:RDF>
```

Fig. 5.1 A sample of SenticNet

Table 5.1 A sample of SenticNet data

A lot	+0.258	Abhorrent	−0.443
A lot sex	+0.858	Able read	+0.865
A little	+0.032	Able run	+0.775
Abandon	−0.566	Able use	+0.856
Abase	−0.153	Abominably	−0.443
Abash	−0.174	Abominate	−0.391

employed the beta version of SenticNet 3.0. It contains 13,741 concepts, of which 7,626 are multi-word expressions, e.g., prevent pregnancy, high pay job, feel happy. Of the concepts in SenticNet, 6,452 are found in WordNet 3.0 and 7,289 are not. Of the latter, most are multi-word concepts such as access internet or make mistake, except for 82 single-word concepts, such as against or telemarketer.

The resource is distributed in RDF XML format (Fig. 5.1) and it is also accessible through an API. The first 20 SenticNet concepts in alphabetical order along with their corresponding polarities are shown in Table 5.1.

The distribution of the polarity values in the dictionary is given in Fig. 1.2. In this figure, each bar corresponds to the number of concepts in SenticNet 3.0 that have the given digit after the dot: for example, +0.0 corresponds to the interval from +0.000 to +0.099.

The distribution of the polarity intensity irrespective of its sign (positive or negative) is given in Figs. 5.2 and 5.3. Again, each bar corresponds to the number of concepts in SenticNet that have the given digit after the dot in their intensity: for example, 0.0 corresponds to the interval from 0.000 to 0.099. As it can be observed from Fig. 5.2, the majority of concepts have low intensity, and for intensities lower than 0.5, the lower the intensity the greater number of concepts possessing it.This gives the idea that concepts with very low polarity intensity were not included in the dataset, and that concepts not listed in the dictionary can be assumed to have almost null polarity (Fig. 5.4).

It can also be observed that the words with a polarity intensity of about 0.5 form the smallest group. This might be an artifact of the labeling procedure or indeed a real linguistic phenomenon. Either way it does suggest that this is a natural division point between high and low polarity: the concepts with high polarity are those that have significant emotional charge.

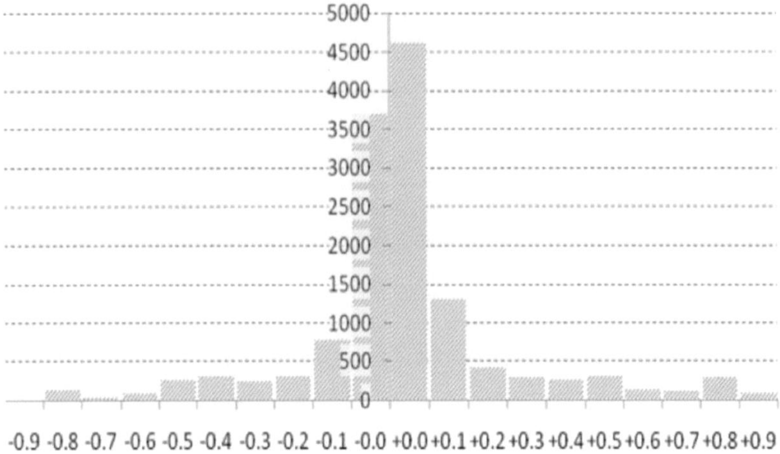

Fig. 5.2 Distribution of polarity intensity in SenticNet

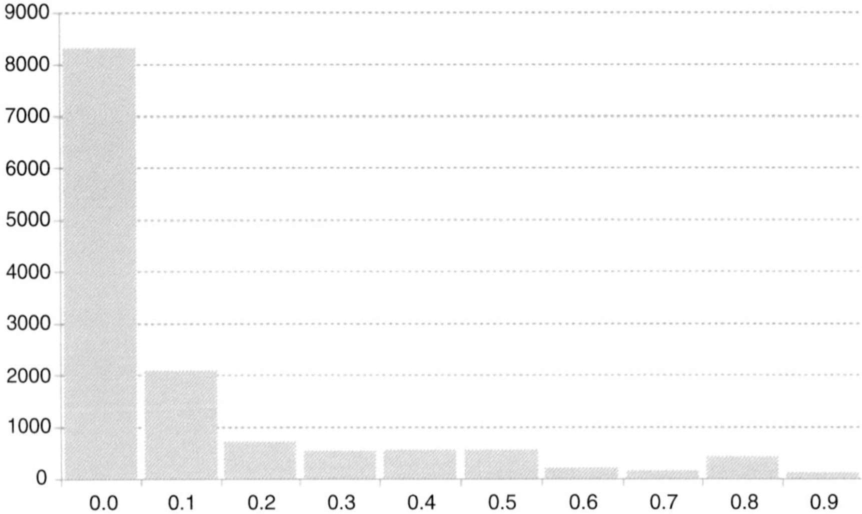

Fig. 5.3 Distribution of polarity intensity in SenticNet

5.2.2 *WNA Emotion Lists*

As an inventory of target labels and a source of training examples for the supervised classification, we used the emotion lists provided for the SemEval 2007 task 14: Affective text. According to the organizers of this task, the lists were extracted from WNA [365]. There are six lists corresponding to the six basic emotions: anger, fear, disgust, sadness, surprise, and joy. This dataset assigns emotion labels to synsets groups of words or concepts that are synonymous in the corresponding senses: e.g.,

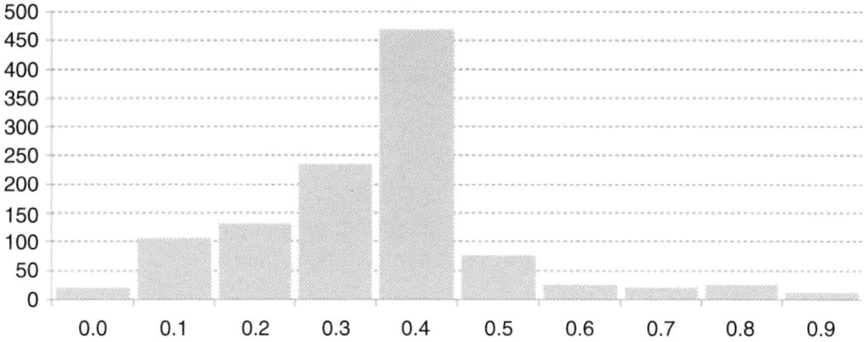

Fig. 5.4 Distribution of polarity intensity of the words from WNA lists in SenticNet

Table 5.2 The number of synsets and concepts by emotion label in WNA lists

	Synsets	Concepts
JOY	204	400
ANGER	115	255
SADNESS	95	202
FEAR	76	147
SURPRISE	27	71
DISGUST	17	53

a synset puppy love, calf love, crush, infatuation is assigned the label JOY. However, we ignored the synonymy information contained in the data and used the labels for individual words or concepts, i.e., puppy love->JOY, calf love->JOY, crush->JOY, infatuation->JOY. Statistics of the synsets and concepts by label is given in Table 5.2. The dataset contains a total of 532 different synsets, of which 2 are assigned two distinct labels each: cliff-hanging, suspenseful, suspensive -> FEAR, JOY and persecute, oppress, harass -> ANGER, SADNESS. Thus, the numbers of synsets in Table 5.2 sum up to 534.

If synsets are broken down into individual concepts (single- or multi-word expressions), the dataset contains 1,113 concepts, of which 15 are assigned two distinct labels each (thus the total of 1128 in Table 5.2). Apart from the words from the two ambiguous synsets, these are words that belong to different unambiguous synsets (in different senses), e.g., awful- FEAR, SURPRISE: when breaking synsets into individual concepts, we lose the sense disambiguation information. Of the obtained concepts, 63 are multi-word expressions, e.g., with hostility or jump for joy; all of them are unambiguous, i.e., are assigned only one label. Only 42 synsets contain multi-word concepts. Of the concepts included in the lists, all but 11 (99.02%) are contained in SenticNet. By considering the emotions JOY and SURPRISE as positive and ANGER, DISGUST, FEAR, and SADNESS as negative, we can assign binary polarity to the listed concepts. The number of concepts is then as follows:

positive 464
negative 643
ambiguous 6

where we count a concept as ambiguous if it has two labels with contradicting polarity, e.g., suspensive -> JOY, FEAR. Of the 1,121 concepts present both in SenticNet and in WNA lists, 5 were found to have ambiguous WNA polarity, and the rest distributed according to the following confusion matrix:

In WNA	Positive	negative
Positive	467	3
Negative	5	646

This shows very good (99.28%) agreement between WNA lists and SenticNet as to the sign of concept polarity. In some cases of disagreement, the polarity assigned by WNA lists seems to be inappropriate, e.g., worry appears under JOY (actually, as a synonym of 'interest'), or stupid under SURPRISE (as a synonym of 'dazed').

5.2.3 *The ISEAR Dataset*

As a source of various features and similarity measures between concepts, we employed the International Survey of Emotion Antecedents and Reactions (ISEAR) dataset [336]. The survey was conducted in the 1990s across 37 countries involving approximately 3,000 respondents. The respondents were instructed to describe a situation or event in which they felt a particular emotion, in the form of a statement – a short text of a couple of sentences (2.37 on average). Here is an example of a complete statement: I had the window open and the door was shut so that the kitten would not go out. My partner came in and started talking about something and I forgot about the window and suddenly I saw the kitten hanging from the window frame. I was rigid with fright till I got hold of her. The full dataset contains 7,666 such statements, which include 18,146 sentences and 449,060 running words. Of the 13,741 concepts contained in SenticNet, 3,312 were found in ISEAR. For these concepts important features were extracted from ISEAR and emotion labels were assigned. Each statement in the ISEAR dataset is supplied with 40 numeric or categorical values, which give various kinds of information on the given situation and the respondent. Thus, the dataset is arranged in a table, with a statement column and 40 data columns. Some of these columns are not informative for our goals, such as the statement ID, the respondent ID, etc. We used only 16 data columns; they are presented in five groups in Table 5.3.

The majority of parameters (except age, gender, religion, country, and emotion) are numerical scores with a small (around 3–4) number of discrete values

Table 5.3 The data columns of ISEAR dataset used in this work

Short name	Description
Background	Background data related to the respondent: age; gender; religion; father's occupation; mother's occupation; country
General	General data related to the emotion felt in the situation described in the statement: intensity; timing; longevity
Physiological	Physiological data: ergotropic arousals; trophotropic arousals, felt change in temperature
Behavioral	Expressive behavior data: movement, non-verbal activity; paralinguistic activity
Emotion	Emotion felt in the situation described in the statement

expressing different degrees of the parameter. For example, the values for ergotropic arousal are: 1 for change in breathing, 2 for heart beating faster, 3 for muscles tensing/trembling, 4 for perspiring/moist hands; and the values for trophotropic arousal are: 1 for lump in throat, 2 for stomach troubles, 3 for crying/sobbing. One of the columns gives the name of the emotion felt. Seven values are used: ANGER, DISGUST, FEAR, GUILT, JOY, SADNESS, and SHAME; in the example above the label was FEAR. This set of seven emotions is slightly different from our target set of six basic emotions used in WNA lists: ANGER, FEAR, DISGUST, SADNESS, SURPRISE, and JOY. A more comprehensive overview of different sets of basic emotions can be found in [252]. We do not directly use the ISEAR emotion labels to assign our target emotion label, but we do use them as one of the feature types for our classification, alongside many other features. The choice of ISEAR as the source of corpus-based information is motivated by the following considerations:

- This corpus is particularly rich in emotion-related words, as compared with more standard corpora used in natural language processing;
- Its statements are consistent with the emotion content. Thus, the "one emotion per discourse" principle, [121], can be applied: if two expressions co-occur in the same statement, then they are related to the same emotion.

In our sample statement the concepts window open, forget, suddenly, hang, rigid with fright are all associated with the same emotion, FEAR. This property makes the ISEAR database particularly suitable for co-occurrence-based emotion similarity measures between concepts.

5.3 Features Used for Classification

The data units that we classified, and to which we assigned the emotion labels (classes) as our final result were concepts: single- or multi-word expressions present in SenticNet. In type-token terms, when gathering information from corpora such as ISEAR, we aggregated the information on multiple occurrences of the concept as a token in the text into one feature vector for the given concept as a type. For our classification we used two kinds of features for the concepts: (1) those based on the data columns directly provided in the ISEAR dataset, and (2) those based on various similarity measures between concepts.

5.3.1 Features Based on ISEAR Data

We used the 16 ISEAR data columns listed in Table 5.3 as independent features. We treated all the features used, with the exception described below, as categorical features. For example, the country column has 16 different numerical codes, so we used 16 different features corresponding to each specific country. As the value, we used 'term frequency': if the concept occurs in the ISEAR dataset 3 times under country code 1 and 5 times under country code 2, then the corresponding part of the feature vector was (..., 3, 5, ...). The values expressing the degree or intensity of various parameters were, for simplicity, treated in the same way; there are around 3–4 discrete values per such data column in the ISEAR dataset. We did not use numeric data types for the values to avoid problems in aggregating (e.g., averaging) values for different occurrences of the same concept. The only exception was the age column. We tested two different approaches: to treat all values (integer number of years) as different categorical values, or to group them in ranges – the latter was found to give better results. According to [271], a psychologically motivated grouping of ages is: 18–23 years (all respondents of the ISEAR datasets were at least 18 years old), 23–28 years, and older than 28 years. So we used these tree intervals as categorical features. This gave us about 100 categorical features, which were used as different dimensions of the feature space.

5.3.2 Features Based on Similarity Measures

Another kind of feature was given by similarity measures between concepts. Given N concepts (N = 3,312 in our case), measuring similarity between pairs of concepts provides N distinct dimensions of the feature vector: the similarity between the given concept and each other concept in the vocabulary. This corresponds to the intuition that if, for two data points, their Euclidian distances to a number of other points are similar, then these two points are probably close to one another. To define these similarity-based features for classification, we used the following 13 similarity measures: 10 lexical resources-based measures (one SenticNet score-based similarity and nine WordNet-based similarity measures) and three co-occurrence-based measures (text distance-based similarity, point-wise mutual information, and emotional affinity) as described below. This contributed 13N dimensions to the feature space. The three co-occurrence-based measures (and in fact some of the WordNet similarity measures that incorporate corpus-based co-occurrence information) are highly correlated but still reflect different granularity levels of the text. Therefore, we included all of them in the feature vectors as independent features.

- SenticNet score-based similarity – The distance D_{SN} between two concepts a, bin SenticNet was defined as $D_{SN}(a, b) = |p(a) - p(b)|$, where p(.) is the polarity specified for the concepts in SenticNet; the similarity was then the inverse of the distance: $Sim_{SN}(a, b) = 1/D_{SN}(a, b)$, or infinity in case of p(a) = p(b). Since all concepts that we considered were present in SenticNet, they had valid SenticNet scores.
- WordNet distance-based similarity – The open-source package WordNet::Similarity [270] based on English WordNet 3.0 was used to measure the WordNet-based distance between concepts. This package provides nine similarity measures based on the analysis of the WordNet hierarchy, glosses, and other data present in WordNet. Namely, we used these following measures:

  1. A simple edge counting approach
  2. Hirst and St-Onge [145]
  3. Leacock and Chodorow [201]
  4. Extended Gloss Overlaps (Banerjee and Pedersen) [268]
  5. Lin [207]
  6. Jiang and Conrath [167]
  7. Resnik [319]
  8. Gloss Vector (Patwardhan et al.) [268]
  9. Wu and Palmer [403]

  In this work, we used all the above nine measures as independent sources of information, each one corresponding to its own N dimensions in the feature vectors. As mentioned in Sect. 5.6.1, not all concepts from SenticNet were present in WordNet 3.0. Of the 3,312 SenticNet concepts that were found in the ISEAR database, 1,436 were directly found in WordNet. Those concepts not found in WordNet, were examined manually and rephrased. For example, if a multi-word concept not present in WordNet, such as make mistake, contained a word that carried its main semantics and was not a stop word, the concept was manually reduced to this main word, mistake in our example. After this, 169 more concepts proved to map to WordNet, giving in total 1,605 concepts with meaningful WordNet pairwise similarity scores. For the remaining 1,124 concepts not mapped to WordNet, their N similarity values to all other concepts were set to random values in the interval [0, 1]. We did not set those values to 0 or some other fixed value because this made all concepts not found in WordNet, very far from other concepts and very similar to each other: indeed, 70 of their features (9N) would be identical. This made them form one large cluster, which deteriorated the final results. In contrast, using random values better expressed the idea of unknown similarity, placing all such concepts in the feature space far from all others and also from each other. All nine similarity scores are defined for specific senses and not for just character strings. For ambiguous concepts, we defined similarity as the maximum similarity overall senses of the first and the second concept.

- ISEAR text distance-based similarity – The positional information of concept tokens in the ISEAR statements was used to measure the ISEAR-based similarity between them. For this, we calculated the average distance between the concepts in the ISEAR dataset statements. Namely, if the tokens a and b occur in a statement S at the positions (in words) $a_1, \ldots, a_n$ and $b_1, \ldots, b_n$, correspondingly, then the distance between a and b in this statement was defined as $D_{ISEAR}(a, b, S) = min(|a_i - b_j|)$, and the distance over the entire ISEAR dataset was defined by averaging over individual statements where both concepts co-occur S_k: $D_{ISEAR}(a, b) = avg D_{ISEAR}(a, b, S_k)$. The similarity was defined as the inverse of the distance: $Sim_{ISEAR}(a, b) = 1/D_{ISEAR}(a, b)$.Note that if the two tokens appear next to each other (as a bigram), then $D_{ISEAR}(a, b, S) = 1$; in particular, $D_{ISEAR}(a, b, S) \geq 1$. If the concepts did not co-occur in any statement, then we considered $Sim_{ISEAR}(a, b)$ to be a random number between 0 and 1.
- Point-wise Mutual Information – The point-wise mutual information (PMI) between concepts measures the degree of co-occurrence between them within a sentence. For concepts a and b, it is defined as

$$Sim_{PMI} = \log \frac{P(a, b)}{P(a)P(b)} \tag{5.1}$$

where P(a) is the probability for a sentence in the corpus to contain a, i.e., the number n(a) of sentences where a occurs, normalized by the total number of sentences in the corpus, and P(a,b) is the probability for a sentence to contain both a and b, i.e., the normalized number n(a,b) of sentences that contain both a and b. Ten concept pairs with the greatest Sim_{PMI} are given in Table 5.4.
- Emotional affinity – We defined the emotional affinity between two concepts a and b in the same way as Sim_{PMI} but at the level of entire statements and not sentences, i.e., P(.) in (1) was defined as the corresponding number of statements

Table 5.4 Top 10 SIM_{PMI} pairs

Concept pair		Affinity
Weekend	December	3 .864
Birthday	Celebration	3.839
Happy	December	3.821
Angry	Friend	3.813
Learn	School student	3.813
Party	Friend	3.813
Graduate	School examination	3.811
Disgust	Behavior	3.807
Money	Important	3.807
Mistake	Realize	3.806

Table 5.5 Top 10 emotional affinity pairs

	Concept pair	Affinity
Weekend	December	3.827
Unemployed	Sad	3.800
Tired of	Headache	3.768
Happy	December	3.755
Worry	Computer virus	3.721
Need relax	Annoyed	3.707
Mistake	Realize	3.691
Disgusting	To person	3.604
Serious	Feel guilty	3.518
Birthday	Thought	3.499

instead of sentences, normalized by the total number of statements. Similar to Table 5.4, the top ten emotional affinity pairs are given in Table 5.5.

While PMI often reflects syntactic association between the concepts – for example, it is high for a verb and its typical object, or for components of a frequent multi-word expression – emotional affinity incorporates a wider notion of relatedness within the same real-world situation, as well as synonymy and rephrasing. Due to our "one emotion per discourse" principle for the ISEAR dataset, the concepts with high emotional affinity tend to be related with the same emotion.

5.4 Fuzzy Clustering

The first step in our process was unsupervised: we did not use the emotion labels known from WNA list. At this step, we grouped together similar concepts. Later on, these clusters were identified with specific emotion labels to construct the six clusters: one for each of the six WNA basic emotions. We considered fuzzy clusters so that a concept could belong to multiple clusters with different degrees of membership. Fuzzy clustering was applied to determine the preliminary estimation for a concept to be related with each specific group; later in the process only one cluster was finally selected for each concept; see Sect. 5.4.1. The features used for the fuzzy clustering were described in Sect. 5.3, and the result of the clustering, for each concept as well as for each of the six groups, was the membership function of a concept in a class defined by the given emotion label. A number ranging between 0 and 1 represented the estimation of the association between the token and a particular emotion.

5.4.1 Fuzzy C-Means Clustering Algorithm

For fuzzy clustering, we used the fuzzy c-means clustering algorithm (Bezdek, 1981) with a modified objective function as described in Sect. 5.4 below. The well-known fuzzy c-means clustering algorithm takes as input, a set of N points $x_1, x_2, \ldots, x_N$ described via their coordinates in a P-dimensional feature space: $x_k = (x_{k1}, x_{k2}, \ldots, x_k)$. As output, it constructs two sets: a set of c centroids $v_1, v_2, \ldots, v_c$ which are points in the same feature space that represent the c clusters found (where c is a given parameter), and a set of $c \times N$ membership values $\mu_{ik}, i = 1, \ldots, c; k = 1, \ldots, N$, which represent the degree of membership of a point x_k in a class c_i. The membership function can be interpreted as the share of a point that a cluster has, so it is assumed that $0 \leqq \mu_{ik} \leqq 1$ and such shares sum up to unity for each point, that is:

$$\sum_{i=1}^{c} \mu_{ik} = 1, k = 1, \ldots, N. \tag{5.2}$$

To find the optimal distribution of points by clusters and optimal placement of the centroids, the algorithm uses an objective function J, which achieves its minimum when the distribution is optimal: $(\mu_0)= arg \min J(\mu, v)$, where $\mu = \mu_{ik}$ and v = v_i represent the sets of the variables to be found, and μ_0, v_0 are the optimal solutions. An expression often used for J is:

$$J_p(\mu, v) = \sum_{i=1}^{c} \sum_{k=1}^{N} \mu_{ik}^{p} \parallel x_k - v_i \parallel^2 \tag{5.3}$$

where the power $p > 1$ is a parameter that controls the degree of fuzziness of the obtained clusters (for our case, p = 2), and

$$\parallel x - y \parallel = \sum_{i=1}^{P} (x_l - y_l)^2 \tag{5.4}$$

is the Euclidean distance in the feature space. The optimal solution of a constraint optimization problem defined by Eqs. 5.2 and 5.3 is given by the following (see Sect. 5.4.2 below for derivation):

$$v_i = \frac{\sum_{k=1}^{N} \mu_{ik}^{p} x_k}{\sum_{k=1}^{N} \mu_{ik}^{p}} \tag{5.5}$$

$$\mu_{ik} = \frac{1}{\sum_{j=1}^{c} (\frac{\|x_k - v_i\|}{\|x_k - v_j\|})^{\frac{2}{p-1}}} \tag{5.6}$$

Now, a stationary point (μ_0, v_0) of the system (5.5), (5.6) is found iteratively using the steps below: (i) Assign random values to all μ_{ik}; the values are normalized to satisfy the constraints (5.2). (ii) Iteratively re-calculate the values for all vi and then all μ_{ik} according to (5.5) and (5.6), respectively; (iii) Stop when the objective function J changes from the previous iteration less than by a small number ϵ (we used $\epsilon = 0.01$).

5.4.2 *Modified Objective Function*

To achieve more compact clusters in which the most similar elements are clustered together, we incorporated an additional term into the original objective function (5.3):

$$J_p(\mu, v) = \sum_{i=1}^{c}\sum_{k=1}^{N} \mu_{ik}^{p}(\| x_k - v_i \|^2 + \rho \sum_{x \in N_k} \| x - v_i \|^2) \tag{5.7}$$

where the parameter ρ is intended to control the effect of the new term (for our case, $\rho = 1$) and N_k is the set constructed in the following way: For each data point x, we identified the nearest centroid

$$v(x) - argmin_i \| V_i - x \| \tag{5.8}$$

(in case of a tie, an arbitrary centroid was chosen); $N_k = x$ was thus set of all data points with the same nearest centroid as x_k. This additional term provides a 'hint' to the algorithm to increase the membership function of a data point in the cluster with the nearest centroid, thus better grouping the similar points together. Our experiments (see Table 5.7) show that this modification of the objective function improved the accuracy of the results. In our implementation, we constructed the sets N_k on the fly while re-calculating the positions of the centroids according to (5.5) (rather its modified version, (5.10), i.e., when re-calculating v_2, we considered in (5.8) the already re-calculated value for v_1. The change of the expression for the objective function required modification of (5.5) and (5.6). For the sake of completeness, a derivation of the modified equations is provided. A note on the discontinuity of $J_p(\mu_0, v_0)$ is also given below. A necessary condition for the optimum of function (5.7), subject to constraint (5.2), to be achieved at a point (μ_0, v_0) is the vanishing of all partial derivatives $\frac{\partial L}{\partial \mu_{ik}}, \frac{\partial L}{\partial v_{im}}$ (v_{im} are coordinates of the centroid v_i in the P-dimensional feature space), and of the Lagrangian of the system:

$$L = \sum_{i=1}^{c}\sum_{k=1}^{N} \mu_{ik}^{p}(\| x_k - v_i \|^2 + \rho \sum_{x \in N_k} \| x - v_i \|^2) + \sum_{k=1}^{N} \lambda_k (1 - \sum_{i=1}^{c} \mu_{ik}) \tag{5.9}$$

The first term of L is $J_p(\mu, v)$ from (5.7), and the second term, according to the method of Lagrange multipliers, is derived from the constraint (5.2), so that the conditions $\frac{\partial L}{\partial \lambda_k} = 0$ reduce to Eq. (5.2). Given (5.7), we have the following:

$$\frac{\partial L}{\partial v_{im}} = \frac{\partial}{\partial v_{im}} \sum_{l=1}^{N} \mu_{ik}^{p}(x_{kl} - v_{il})^2$$

$$= \sum_{k=1}^{N} \mu_{ik}^{p}(-2(x_{km} - v_{im}) + \rho_k \sum_{x \in N_k} -2(x_m - v_{im})) = 0 \quad (5.10)$$

and, taking into account that $\sum_{x \in N_k} v_{I'm} = v_{I'm} \sum_{x \in N_k} 1 = v_{I'm} \mid N_k \mid$, we have:

$$V_{im} = \frac{\sum_{k=1}^{N} \mu_{ik}^{p}(x_{km} + \rho \sum_{x \in N_k} x_m)}{\sum_{k=1}^{N} \mu_{ik}^{p}(1 + \rho \mid N_k \mid)} \quad (5.11)$$

or, in a vector form,

$$V_{im} = \frac{\sum_{k=1}^{N} \mu_{ik}^{p}(x_k + \rho \sum_{x \in N_k} x)}{\sum_{k=1}^{N} \mu_{ik}^{p}(1 + \rho \mid N_k \mid)} \quad (5.12)$$

which substitutes (5.5) for our modified objective function. Similarly,

$$\frac{\partial L}{\partial \mu_{ik}} = p\mu_{ik}^{p} - 1(\| x_k - v_i \|^2 + \rho_k \sum_{x \in N_k} \| x - v_i \|^2) - \lambda_k = 0 \quad (5.13)$$

gives

$$\mu_{ik} = \frac{\lambda_k^{\frac{1}{p-1}}}{p^{\frac{1}{p-1}}(\| x_k - v_i \|^2 + \rho \sum_{x \in N_k} \| x - v_i \|^2)^{\frac{1}{p-1}}} \quad (5.14)$$

which together with (5.2) gives

$$\lambda_k^{\frac{1}{p-1}} = \frac{1}{\sum_{i=1}^{C} \frac{1}{p^{\frac{1}{p-1}}(\|x_k - v_i\|^2 + \rho \sum_{x \in N_k} \|x - v_i\|^2)^{\frac{1}{p-1}}}} \quad (5.15)$$

and thus:

$$\mu_{ik} = \frac{1}{\sum_{j=1}^{C} \left(\frac{\|x_k - v_i\|^2 + \rho \sum_{x \in N_k} \|x - v_i\|^2}{\|x_k - v_j\|^2 + \rho \sum_{x \in N_k} \|x - v_j\|^2}\right)^{\frac{1}{p-1}}} \quad (5.16)$$

which substitutes (5.6) for our modified objective function. If $\rho = 0$, then (5.10) and (5.11) reduce to (5.5) and (5.6), respectively. This derivation only applies to the regions where is differentiable. In fact it is not everywhere continuous, since the sets Nk change sharply when the data points jump from one N_k to another as v_i changes. Therefore, our analytic solution will only find a local optimum within a region of continuity of $J_p(\mu, v)$. This is, however, common practice, as the original solution (5.5), (5.6) does not guarantee a global optimum, either. In such cases, estimation of the performance of the resulting iterative algorithm is left to empirical evaluation, which is given in Sect. 5.7.

5.4.3 Mapping Fuzzy Classes to Emotion Labels

Following the completion of the unsupervised fuzzy clustering process, we identified which of the $c = 6$ classes appropriately corresponded to one of the six emotion labels. For this, we employed a simple supervised approach. First, we converted the fuzzy clustering into hard clustering; in our implementation we chose for each data point x_k a cluster $C(x_k) = argmax_i \mu_{ik}$ to which it was most strongly associated (in case of a tie, an arbitrary class was chosen). Then the emotion label for each hard cluster was chosen by majority voting. Namely, considering all concepts in the cluster that happened to be in WNA emotion lists and thus had known emotion labels, we calculated the number of times each label occurred (for those concepts that had more than one emotion label in the emotion lists, all their labels were counted), and then selected a label that occurred the highest number of times. This procedure did not guarantee that there be six hard clusters, or that the majority voting would not result in a tie, nor that two clusters would share the same emotion label. In the latter case, some labels would not be assigned to any cluster. However, such low probable events did not occur in our experiments. Moreover, correctness of the obtained mapping of the classes to emotion labels is confirmed by the fact that we obtained over 90% accuracy in our final results, which would not be possible with incorrectly mapped labels.

5.5 Hard Clustering

While several concepts appear in more than one WNA list (for example, harass is listed under SADNESS and ANGER, suspensive under JOY and FEAR), most of them have only one emotion label. Predicting whether a word is ambiguous, is out of the scope of the present work, and it cannot be said with certainty if such ambiguity might be the result of errors in WNA lists. Therefore, to simplify things, we chose to assign only one emotion label to each concept. In our evaluation, we consider a label to be assigned correctly if the WNA lists assigned this label to the concept – even if it additionally assigns another label to it. There are however, too few cases of double labels present in the lists for this to significantly alter our results. A two-step process was used for choosing a single class for each token classification.

5.5.1 *Reduction of the Confusion Set*

Given a parameter K discussed below, for each concept, we chose top K classes, that is, K classes for which the fuzzy clustering assigned this concept the highest value of the membership function. For example, if K = 2 and the six membership functions for a given concept were 0.1, 0.8, 0.3, 0.4, 0.6, 0.2, then classes 2 and 4 were selected and the classes 1, 3, 5, and 6 discarded as having lower membership functions for this concept than the K = 2 selected classes. The hard clustering technique used afterwards was only allowed to choose between those top K labels pre-selected for a given concept. In case of K = 1, only one class per concept is chosen, and thus no further processing is needed: in this case, the final result is given by choosing the greatest membership function obtained from the fuzzy clustering. In case of K = 7, there is no reduction of the confusion set, i.e., the reduction step is in effect omitted. In case of K = 2 or K = 3 – the values we experimented with – the confusion set is reduced to 2 or 3 options, correspondingly. We show in Sect. 5.7.1 that reducing the confusion set to only two candidates increases the accuracy.

5.5.2 *Final Hard Classification*

Given the K options left after reducing the confusion set, we trained separate classifiers for each of the $\binom{6}{k}$ possible combinations of K emotion labels: for example, with K = 2, a separate classifier was trained for choosing between FEAR and DISGUST, another one to choose between FEAR and SADNESS, etc., which gives $6 \times 5/2 = 15$ different classifiers. To assign a unique label to a concept, the K emotion labels for it were selected as explained above, e.g., FEAR and SADNESS, and then the corresponding classifier was used. To train a classifier for a given set of labels, we used only those concepts that had any of the corresponding labels in the WNA lists. For example, to train a classifier for the confusion set FEAR, SADNESS, we used all concepts extracted from the ISEAR corpus (for which, therefore, we had their feature vectors) that were present in WNA lists with either the label FEAR or the label SADNESS. (The few concepts with double labels, such as harass, were excluded from the training data for those confusion sets that contained both labels.) As features, we used the same feature vectors that were used for the fuzzy clustering, extended by six extra dimensions: the membership values generated by the fuzzy classifier for the six emotion labels, except the experiments where the fuzzy clustering was not used. As classes, the confusion set of K selected labels was used for each classifier; in case of $K = 2$, the classification was binary. As a hard clustering algorithm, we used the SVM framework. Specifically, we used the libsvm library of the WEKA toolset, which, for the case of $K > 2$, provides an implementation of a multiclass SVM. As a result, we obtained one emotion label for each concept in the dataset.

5.6 Implementation

While in Sects. 5.1.1, 5.1.2, 5.2, 5.2.1, 5.2.2, 5.2.3, 5.3, 5.3.1, 5.3.2, 5.4, 5.4.1, 5.4.2, 5.4.3, and 5.5 we have described an abstract method applicable to any datasets, in this section we give the details of the specific dataset to which we applied the method described above and present examples of the resulting labels.

5.6.1 Data Preparation

We used a number of standard pre-processing techniques, such as tokenizing and lemmatizing, as described below. For this, we used the tools provided by Rapidminer's text plug-in, except for lemmatizing (a lemmatizer differs from a stemmer in that it provides a complete form: for example, for feet, it provides foot), for which we used the WordNet lemmatizer [234]. For each SenticNet concept, we identified all its occurrences in the text of the ISEAR statements. All words in the statements were lemmatized before matching, because the concepts could appear in the text in a different form, e.g., made mistake in the text vs. make mistake in the vocabulary (this also generated some small number of false matches). For multi-word SenticNet concepts, such as after summer, to person, etc., we allowed any number of stop-words to appear in the position of the space, so that, in SenticNet, they were readily matched, e.g., to a person or to the person in the text. Total of 3,312 SenticNet concepts appear at least once in the ISEAR dataset. Only these concepts participated in further processing and were finally assigned the emotion labels. For each occurrence, we extracted the corresponding data fields from the ISEAR dataset, and the data for multiple occurrences of the same concept in the corpus (as a token) were aggregated in a feature vector for that concept (as a type). Thus, this gave us a dataset with a total of 3,312 feature vectors. All 3,312 concepts participated in the unsupervised fuzzy clustering phase, though not all of them participated in the supervised final hard clustering; see Sect. 5.5. Consider the following example: When I make mistake that I has accused someone else of, this is obvious to the person. The concepts found in this example and their corresponding emotion labels assigned by our classifier are make mistake (sadness), obvious (joy), to person (surprise). For the example I am sad when friends try to put me down or hurt me, the corresponding concepts and emotion labels are sad (sadness), friend (joy), put down (sadness), hurt (fear).

5.7 Direct Evaluation of the Assigned Emotion Labels

For evaluation, we used a standard tenfold cross-validation procedure. Namely, we excluded from training a tenth part of WNA emotion labels. Then we normally constructed the resource using this reduced WNA set. In the constructed resource, we compared the labels assigned to the concepts that were temporarily excluded from WNA (but left in SenticNet, so that they were automatically assigned some labels) with the true labels present for these concepts (but unavailable during training); percent of coincidence gave accuracy. This accuracy was averaged by ten experiments where one of ten different tenth parts of WNA was excluded from training and used for evaluation. We tested several classifiers for the final classification task. Of them, SVM produced the best result, with 92.15% accuracy; see Table 5.6. Apart from different classifiers, we compared the effect of selection of different parameters and different subsets of features, in order to study which features are most important for assigning the emotion labels. Below we first describe these experiments in detail.

5.7.1 Impact of the Fuzzy Clustering and Hard Classification

We compared the results of the pipeline of fuzzy clustering and hard classification as described above versus directly applying hard classification without the fuzzy clustering phase; in the latter case, the hard classifier did not have access to the extra features obtained by the fuzzy classifier. The results are presented in Sections A and B below. We also experimented with different values of K introduced in Sect. 5.7.1: the size of the confusion set after reduction based on the result of fuzzy clustering. Namely:

- K = 1 means that the final classification directly results from the fuzzy clustering and no further hard clustering is necessary;
- K = 2 means that the final hard classification is based on binary choices between the two top labels for each concept;
- K = 3 means that reduction of the confusion set for the hard classification to a choice of three options;

Table 5.6 Impact of the selection of most likely fuzzy cluster

K	Accuracy
K = 1: Fuzzy clustering only, no SVM	67.45%
K = 2: SVM applied to top two fuzzy clusters	92.15%
K = 3: SVM applied to top three fuzzy clusters	83.41%
K = 6: no reduction of confusion set	65.43%

- K = 6 means no reduction of the confusion set. However, it is not the same as not using the fuzzy clustering phase at all as described above, because the fuzzy clustering results are still used as additional features for the hard classifier.

In addition, we conducted experiments by SVM alone without prior fuzzy clustering. We found that the fuzzy clustering step does help SVM to achieve better classification. We show that SVM performed best when for each concept we selected two (K = 2) clusters with the highest membership values. As mentioned earlier, this selection reduces the size of the confusion set for the SVM-based classification. Table 5.6 presents a comparison between respectively: reduction to two top classes, reduction to three top classes, and no reduction of the confusion set (K = 6). In these experiments, all features were used (later in this section we will study the impact of different feature combinations on the classification; see Table 5.7 below). In case of no reduction of the confusion set, the results of the fuzzy clustering step were still used by the SVM classifier in the form of additional features: specifically a vector of the membership functions of the given concept in all six clusters. When we chose only one highest membership value, i.e., the strongest cluster as the final classification result for the given concept (K = 1), then no SVM step was needed. The results for this case with different feature combinations are shown in Table 5.7 in the leftmost numerical column. In this table, WordNet stands for the set of WordNet-based features, SenticNet stands for SenticNet similarity feature, Lexical stands for other lexical features, Background, General, Physiological, and Behavioral stands for various ISEAR data-based features (Table 5.3 and Sect. 5.3.1), and Membership stands for the vector of the fuzzy clustering results: membership functions in each of the six clusters. The figures are presented based on tenfold cross-validation.

Table 5.7 shows that using the fuzzy membership vector as a feature for SVM increased accuracy. This fuzzy membership feature, along with all other features, gave the highest accuracy of 92.15% on the test set. As training and test data, we used the intersection between the sets of concepts found in the WNA lists (for which we had the gold standard emotion labels) and those 3,312 SenticNet concepts found in the ISEAR texts (for which we had ISEAR specific features); this intersection consisted of 1,202 concepts. Specifically, the system was evaluated on the 362 WNA concepts from the six WNA lists.

5.7.2 *Impact of Psychological Factors*

We experimented with different combinations of the psychological features extracted from the statements of the ISEAR dataset. The impact of the psychological features is demonstrated in Table 5.6. It can be observed that all ISEAR data-based features improved the classification accuracy, and the best accuracy was obtained with all the features. The individual ISEAR data-based features results based on the general and physiological variables show interesting insights into the effect of the features from the perspective of emotion. For example, low intensity for emotion

Table 5.7 Precision with different feature combinations and different classifier combinations

Wordnet	SenticNet	Lexical	Background	General	Physiological	Behavioral	Membership	Fuzzy only K = 1	SVM only	Fuzzy + SVM
+								47.66%	50.59%	52.02%
	+							25.27%	29.07%	32.23%
+	+							56.64%	57.77%	61.21%
		+						24.12%	27.87%	31.09%
	+	+						25.23%	28.21%	33.01%
+		+						41.32%	43.09%	45.15%
+	+	+						58.11%	59.27%	63.50%
+	+	+	+					69.78%	73.23%	78.26%
+	+	+		+				63.18%	65.12%	66.04%
+	+	+			+			59.17%	60.22%	65.77%
+	+	+				+		55.34%	58.00%	61.10%
+	+	+	+	+				61.02%	61.12%	62.02%
+	+	+	+		+			60.15%	60.24%	61.00%
+	+	+	+			+		58.12%	58.45%	58.75%
+	+	+		+	+			55.77%	56.34%	58.14%
+	+	+		+		+		65.35%	66.12%	69.04%
+	+	+	+	+	+	+		80.11%	85.12%	88.85%
+	+	+	+	+	+	+	+	83.41%	-	92.15%

classes of shame and guilt and high intensity for joy, fear, and sadness [79]. It should be noted that [79] did not use any other type of psychological variables. In contrast, we use the features related to background variables and expressive behavior in addition to the general and psychological variables. One can observe that the expressive behavior contributes less as compared with other groups of variables. Furthermore, we evaluated our results using the polarity sign. We considered anger, fear, sadness, and disgust as negative, and joy and surprise as positive emotions. This allowed us to evaluate the method on all 3,312 concepts against SenticNet data, because the polarity sign (positive or negative emotion) of all concepts present in SenticNet can be inferred from the SenticNet data. This evaluation gave 96.11% precision and 92.24% recall, which is considered to be very satisfactory result.

5.7.3 Error Analysis

We observed a few typical situations leading to errors. One of them is detection of emotion words that are present in the same ISEAR statement but in two separate sentences. The lexical affinity between such remote occurrences does not always imply similarity between their lexical patterns, and therefore sometimes leads to wrong results. In fuzzy c-means clustering, some concepts, such as humiliate, have very similar membership values in more than one cluster (in this case, those corresponding to FEAR and SADNESS). And, sometimes the SVM classifier chose the correct final emotion class for such a concept (SADNESS in this case). However, for certain some ambiguous concepts such as faint, sick, humble, etc., the SVM classifier assigned incorrect emotion label.

5.8 Construction of Emosenticspace

Cambria et al. [43] have combined WNA with ConceptNet [364], a large publicly available common-sense knowledge base that contains simple statements such as "you use a spoon for eating" or "a book is made of paper". The combined resource, termed AffectiveSpace, contains both types of information: semantics and world knowledge on the one hand, and sentics on the other hand, which makes it a powerful resource for reasoning about real-world situations and behavior involving emotions. Since our ESN is a superset of WNA, we extended AffectiveSpace to a combination of ConceptNet with ESN instead of WNA and, hence, obtain EmoSenticSpace. This extended resource contains the same semantic and real-world information but much more sentic information, since more words in it have affective labels.

5.8.1 Building EmoSenticSpace

To build EmoSenticSpace, we followed the procedure proposed by Cambria et al. [43], which is briefly described below. ConceptNet is represented in the form of a labeled direct graph, with nodes being concepts such as, for example, spoon, eating, book, paper, and arcs being relations such as UsedFor (spoon-UsedFor ->eating) and MadeOf (book - MadeOf ->paper). Technically, a graph can be thought of as a matrix. To perform inference on multiple matrices, blending is the most widely used technique. It allows multiple matrices to be combined in a single matrix, basing on the overlap between these matrices. The new matrix is rich in information and contains much of the information shared by the two original matrices. By means of the singular value decomposition on the new matrix, new connections are formed in source matrices based on the shared information and overlap between them. This method enables creation of a new resource, which is a combination of multiple resources representing different kinds of knowledge. In order to build a suitable knowledge base for affective reasoning, we applied the blending technique to ConceptNet and ESN. First, we represented ESN as a directed graph, similarly to ConceptNet. For example, the concept birthday party has the associated emotion joy; we considered birthday party and joy as two nodes, and added an assertion HasProperty on the edge directed from the node birthday party to the node joy. Next, we converted the two graphs, ConceptNet and ESN, to sparse matrices to blend them. After blending the two matrices, we performed the Truncated Singular Value Decomposition (TSVD) on the resulting matrix, to discard those components that represent relatively small variations in the data. We kept only 100 components of the blended matrix to obtain a good approximation of the original matrix. The number 100 was selected empirically: it gave the best results.

5.8.2 Features Used in the Applications

Features Based on ESN's Emotion Labels and EmoSenticSpace In order to demonstrate the effectiveness of both resources on such tasks as sentiment analysis, emotion detection, and personality detection, we used them to extract the following features from documents: Features based on ESN. For each concept found in the document, we obtained its corresponding affective label from ESN, if any. We aggregated the individual concept labels into a six-dimensional vector for the document by counting the frequency of occurrence of each label in the document: say, if the document had five concepts labeled as JOY, the coordinate in the document vector corresponding to JOY was set to 5. This gave us a six-dimensional integer-valued feature vector for each text document. Features based on AffectiveSpace and EmoSenticSpace. For each concept found in the document, we extracted its 100-dimensional feature vector from EmoSenticSpace. We aggregated the individual concept vectors into one document vector by coordinate-wise summation: $(x_i) = (\sum_{j=0}^{N} x_{ij})$; where (x_i) is the i-th coordinate of the document's feature vector, (x_{ij}) is the i-th coordinate of the j-th concept's vector of this document, and N is the number of concepts found in the document. This gave us a 100-dimensional real-valued feature vector for each text document. For comparison, we also computed these 100 features using AffectiveSpace instead of EmoSenticSpace. In the experiments described in the next section, we observed that the 100-dimensional EmoSenticSpace feature vectors gave very good results. Six-dimensional ESN feature vectors performed very poor. However, combined 106-dimensional feature vectors showed very slight improvement over the 100-dimensional EmoSenticSpace vectors. Below we report the results obtained with these 106 dimenstional feature vectors.

5.8.3 *Other Features*

For better results, we used a few other features for the documents in our experiments. For simplicity and given that all three tasks we considered have a common cognitive nature, we used the same set of features for all these tasks. These additional features are as follows:

- Sentic feature: the polarity scores of each concept in the text, obtained from ESN, were summed up to produce a single scalar feature.
- Part of speech features: three features defined as the number of adjectives, adverbs, and nouns in the text.
- Modification feature: the number of sentences in the text that have a word modified by a noun, adjective, or adverb in its dependency tree (we used the Stanford dependency parser).
- Negation feature: the number of sentences in the text that contain negation. The latter feature is important because the negation can invert the polarity of the sentence.

5.9 Performance on Applications

We used ESN and EmoSenticSpace in a number of affective applications. We found that these resources give significantly higher accuracy than that reported so far in the literature for these applications, specifically, sentiment analysis, emotion detection, and personality detection. Identifying emotions, sentiment, and personality in text is a challenging task because of the ambiguity of words in the text, complexity of meaning, and interplay of various factors such as irony, politeness, writing style, as well as variability of language from person to person and from culture to culture. Surprisingly, the same feature set, namely, the 106 features described in the previous section, worked excellently for all three tasks. For comparison, we also give the results obtained with 100-dimensional feature vectors extracted from the original AffectiveSpace. In all three experiments, we used tenfold cross-validation for evaluation.

5.9.1 *Sentiment Analysis of Text*

For experiments on detecting positive and negative sentiment in texts, we used Stanford Twitter dataset [127]. This resource gives binary polarity labels (POSITIVE/NEGATIVE) for a large number of tweets. We cast this task as a binary classification task. For sentiment analysis experiment, this was binary classification. We report the results obtained with the SVM as the classifier. The accuracy obtained with SVM was superior to that obtained with other supervised classifiers we tried, such as other state-of-the-art Artificial Neural Network (ANN) and Naïve Bayes classifiers. Table 5.8 shows the experimental results and presents a comparison between our approach and the highest state of the art accuracy reported so far in literature. While the original AffectiveSpace performed quite poorly, our EmoSenticSpace outperformed the best state of the art approach.

5.9.2 *Emotion Detection from Text*

As a dataset for the emotion detection experiment, we used the ISEAR dataset. We cast the task as a seven-way classification, where the seven classes were the emotion labels used in the ISEAR dataset: ANGER, DISGUST, FEAR, GUILT, JOY, SADNESS, and SHAME. Note that these classes differ from the six emotion labels that we used in ESN. Again, SVM was found to give comparatively better accuracy than other supervised classifiers. Table 5.9 shows a very significant improvement achieved by our approach over the highest state of the art accuracy result we are aware of. As can be seen from Table 5.9, despite the original AffectiveSpace significantly outperforming the best state of the art approach, EmoSenticSpace was found to perform even much better.

Table 5.8 Sentiment analysis on Stanford Twitter dataset

Method	Accuracy
ESN and AffectiveSpace	72.35%
Go et al. (2009) [127]	83.10%
ESN and EmoSenticSpace	85.05%

Table 5.9 Emotion detection on the ISEAR dataset

Method	Accuracy
Kim et al. (2010) [188]	52.80%
ESN & AffectiveSpace	61.20%
ESN & EmoSenticSpace	67.25%

Table 5.10 Personality detection on the essays dataset for personality detection

Method	O	C	E	A	N
Mohammad and Kiritchenko [236]	60.4%	56.4%	54.6%	54.0%	55.7%
Mairesse et al. [220]	62.1%	55.2%	54.9%	55.7%	57.3%
ESN and EmoSenticSpace	**66.1%**	**63.3%**	**63.4%**	**61.5%**	**63.7%**

5.9.3 Personality Recognition from Text

For experiments on detection personality from text, we used five-way classification according to the five personality traits described by [225], which are openness, conscientiousness, extraversion, agreeableness, and neuroticism, sometimes abbreviated as OCEAN, by their first letters. We used the dataset provided by [220]. This dataset provides student essays along with a binary vector of personality traits of the author of each essay. The vector indicates the presence or absence of each of the five OCEAN traits, such as openness: yes, conscientiousness: no, extraversion: yes, etc. We cast the task as five independent binary classification tasks. In this case, SVM only slightly outperformed other supervised classifiers in terms of accuracy. Table 5.10 shows the results of this experiment and a comparison with two best state of the art approaches we are aware of; the letters in the header correspond to the OCEAN label set. In these experiments, the original AffectiveSpace's performance was only slightly lower that of EmoSenticSpace. More details on these experiments can be found in [296]. Therefore, in all our experiments the resources that we developed, ESN and EmoSenticSpace, significantly outperformed existing state of the art techniques.

5.10 Summary of Lexical Resources and Features Used

We built the EmoSenticSpace resource by first extending the emotion labels from WNA to SenticNet, and then blending the obtained resource, which we called ESN, with ConceptNet. Below we remind the reader the various resources mentioned in this work and their relationships with each other. In addition, the first of the two resources was obtained via supervised machine learning. As in any typical machine learning setting, our work consisted in mapping a set of data points (in our case vocabulary – a list of words and multiword expressions denoting concepts), to some categories (in our case, emotion labels) via a set of features (extracted from lexical resources), using a source of ground truth labels both for training (along with

Table 5.11 Summary of vocabulary, features, and training categories used in development of ESN, along with lexical resources that provided them

Features	SenticNet	WNA	ISEAR data	ISEAR text	WordNet
Vocabulary	+				
Background			+	+	
General			+	+	
Physiological			+	+	
Behavioral			+	+	
Emotion			+	+	
SenticNet score-based similarity	+				
WordNet distance-based similarity					+
ISEAR text distance-based similarity				+	
Point-wise mutual information				+	
Emotional affinity				+	
Mapping classes to labels (training)		+			
Ground truth labels (testing)		+			

unlabelled training data) and testing by tenfold cross-validation. In this section, we summarize our use of lexical resources to extract the vocabulary, the features, and the ground truth labels for building ESN.

The lexical resources and the features we used for the development of ESN are summarized in Table 5.11. Specifically, for the development of ESN we used the following four lexical resources:

- SenticNet is a dictionary that provides average polarity for words and concepts. We used it to extract both vocabulary (which was the complete vocabulary of this dictionary) and the SenticNet score-based similarity feature, which measured how similar are the polarity values associated with two words.
- WNA Emotion Lists are a dictionary that provides emotion labels for a small number of words and concepts. We used it as a source of both training instances (to associate fuzzy classes with specific labels, as described in Sect. 5.2.2) and ground truth testing instances.
- The ISEAR Dataset is a collection of short texts (statements) describing a situation in which an emotion was felt, along with 40 numerical or categorical data items for each statement specifying the emotion and describing the person that felt the emotion and other circumstances accompanying the situation. We used many of those data items as features for words and concepts from our target vocabulary that occurred in the statements. In addition, we used the text of the statements (ignoring the data items) as a text corpus for learning various co-occurrence measures between words that occur in these texts.

Table 5.12 Summary of the use of lexical resources as sources of features and training categories used in our applications

	Sources of features					Testing and training datasets		
Application of ESN	ESN							
	Labels	Scores	Concept-Net	Affective-Space	EmoSenticSpace	Twitter dataset	ISEAR dataset	Personality dataset
Building EmoSenticSpace	+		+					
Sentiment analysis	+	+		±	+	+		
Emotion detection	+	+		±	+		+	
Personality Recognition	+	+			+			+

For development of ESN, we used the following data points, features, and categories extracted from these lexical resources:

1. Data points (vocabulary of concepts to be assigned labels) were words and concepts that we aimed to map to final emotion labels. We extracted them from SenticNet's vocabulary.
2. Features of those data points (words or concepts) were as follows:
   a. Features accompanying a situation when an emotion was felt: the background of the person who felt it, general characteristics of the situation, psychological arousals, behavioral data of the person, and the name of the emotion described by the text containing the word or concept. All these features were extracted from the ISEAR dataset.
   b. Similarity scores: the closeness of polarity (extracted from SenticNet), numerous variants of calculation of the distance in WordNet graph, and co-occurrence frequency at different granularity (as observed in the ISEAR texts). Each similarity score contributed N (the size of the vocabulary) individual features to the feature vectors of concepts: the similarity of the given concept to every other concept in the vocabulary.
3. The target categories, or emotion labels to be assigned to the concepts, were used both for training and as ground truth for testing. They were extracted from the WordNet Affect lists.

The lexical resources we used in the three applications of EmoSenticSpace (and thus indirectly of ESN) are summarized in Table 5.12. Specifically, in our applications we used the following seven lexical resources:

1. **EmoSenticNet (ESN)** is a dictionary built in this work, which provides average polarity and emotion labels for a large number of concepts. First, we used it to extract emotion information to be blended with ConceptNet to obtain EmoSenticSpace. Then, we used it to extract the polarity score and emotion

category features for our three applications. ConceptNet is a semantic network dictionary that provides relations between concepts, such as *spoon -> "UsedFor" -> eating*. We used it to build our combined resource, EmoSenticSpace.
2. **AffectiveSpace** is a resource similar to our EmoSenticSpace but built with WNA instead of our much larger ESN. We do not use this resource in our proposed method, but compare the use of our EmoSenticSpace with the use of AffectiveSpace to show the advantages of our resource.
3. **EmoSenticSpace** is the second lexical resource that we built. We obtained it by blending our ESN with ConceptNet and reducing the resulting graph to a matrix of concepts by features. For each concept from ESN or ConceptNet's vocabulary, it gives 100 unnamed features, which result from an algebraic transformation of a much larger graph obtained from ConceptNet and ESN. This resource is similar to AffectiveSpace but much richer in information.
4. **Stanford Twitter dataset** is a set of Twitter samples supplied with binary polarity labels (POSITIVE/NEGATIVE). We used it as a training and testing set for our sentiment analysis application.
5. **The ISEAR Dataset** described above is a collection of short texts describing a situation in which a specific emotion was felt; this emotion is specified for each text. We used it as a training and testing set for our emotion detection application.
6. **Personality dataset** [220] is a collection of student essays along with indication of whether each of the five personality traits is present or absent in the author's personality profile. We used it as a training and testing set for our personality recognition application.

For our applications, we used the following data points, features, and categories extracted from these lexical resources:

1. Data points were documents from the corresponding dataset for each of the three tasks that we aimed to map to the labels specified in this dataset.
2. Features of those data points (documents) were as follows:
    a. Sentic and common-sense features: the count of each emotion label for all concepts in the document (6 integer features), determined with ESN; the coordinate-wise sum of the EmoSenticSpace features for all concepts in the document (100 numeric features); and the sum of the polarity scores for all concepts in the document (1 numeric feature), determined with ESN.
    b. Syntactic features: the count of adjectives, adverbs, and nouns in the document (3 integer features) and the counts of sentences in the document with modification constructions and with negation construction (2 integer features).
    c. The target categories for classification were those specified by the resource used for each application:
        i. For the sentiment analysis application, POSITIVE vs. NEGATIVE polarity.
        ii. For the emotion detection application, the seven ISEAR emotion labels.

d. For the personality recognition application, five independent binary features corresponding to the OCEAN set; that is, personality recognition application consisted in five independent binary classification tasks.

In Table 5.12, the sign +/− stands for the use of AffectiveSpace only for comparison with the use of EmoSenticSpace, but not as part of our proposed methods.

5.11 Conclusion

In this chapter, an empirical method for assigning emotion orientation labels to the concepts of SenticNet is proposed, or, in other words, for expanding WNA to a much larger vocabulary of concepts present in SenticNet. Thus, the resulting resource, ESN, can be thought of as augmenting SenticNet with emotion labels, or as an extension of WNA to a much larger vocabulary. The resource is publicly available. Given that emotion orientations are fuzzy in nature, we used the fuzzy c-means clustering algorithm to initially cluster the concepts into fuzzy clusters. The results of this fuzzy clustering were used in two ways to facilitate the final hard categorization into six emotion labels: first, to reduce the confusion set to the top two labels, and second, as additional six features used by the hard classifier. For the hard classification stage we employed a state-of-the-art SVM. We exploit some novel features, such as point-wise mutual information and similarity scores, to determine emotional affinity between two concepts. We also introduced effective modifications to the conventional fitness function for fuzzy clustering. Further, we investigated the impact of different features, including psychological features, and the performance of our method with different feature combinations. The method showed 92.15% accuracy on the best combination of features. Since ESN is an extension of SenticNet and of WNA, it can be used instead of these resources to achieve better results. Consequently, we also employed it to improve the AffectiveSpace resource by substituting the WNA. The constructed resource, EmoSenticSpace, in combination with the direct use of ESN, allowed for achieving a significant improvement over the state-of-the-art on such tasks as sentiment analysis, emotion detection, and personality detection in texts. There are a number of directions for future work. One is the incorporation of syntactic information. Another is to develop solutions for typical error cases. The textual clues related to psychology may be included to improve the performance of the method. In our current work, we were able to assign emotion labels only to 3,312 concepts of SenticNet that appeared in the ISEAR corpus. We plan to use additional monolingual or multilingual [357] lexical resources to be able to assign emotion labels to all concepts of SenticNet, as well as to construct emotion lexicon for other languages [6, 80, 388]. We plan to apply our resource to opinion mining tasks, as well as cognitive applications such as identifying emotion in music or mining opinion in narrative financial disclosures. We will also employ ESN in tasks at which SenticNet or WNA have been previously applied, to evaluate comparative performance improvements. In this work, we reduced fuzzy clusters to a single label for each concept. However, emotion labeling should be considered as a multi-label problem, because a word can either invoke different emotions in different contexts, or invoke more than one emotion at the same time. Thus assigning only one emotion label to each concept may be misleading. For example, the concept succumb invokes sadness and, depending on the context, shame or fear. If we only assign it shame, we lose the additional information that this concept invokes sadness, and will report misleading information if, in a particular context, the concept invokes fear and not shame (succumb to temptation vs. succumb to disease). In the future, we will explore the possibility to use fuzzy clustering results directly or to assign to concepts multiple labels, probably weighted, and preferably anchored in the context, which can be useful for multi-dimensional opinion mining in systems, such as movie recommender. This would also open the way to contextual polarity classification, where the same textual content can be presented with different emotional slants.

Chapter 6
Sentic Patterns: Sentiment Data Flow Analysis by Means of Dynamic Linguistic Patterns

Abstract In this chapter, we introduce a novel paradigm to concept-level sentiment analysis that merges linguistics, common-sense computing, and machine learning for improving the accuracy of tasks such as polarity detection. By allowing sentiments to flow from concept to concept based on the dependency relation of the input sentence, in particular, we achieve a better understanding of the contextual role of each concept within the sentence and, hence, obtain a polarity detection engine that outperforms state-of-the-art statistical methods.

Keywords Semantic parsing · Linguistic patterns · Machine learning · Polarity detection · Ensemble classification

6.1 Introduction

The Bag of Concepts (BoC) model can represent the semantics associated with a natural language sentence much better than BoW. For example, a concept such as `cloud computing` would be split into two separate words, disrupting the semantics of the input sentence (in which, for example, the word `cloud` could wrongly activate concepts related to `weather`). The BoC model, however, would not be able to correctly infer the polarity of a sentence such as "the phone is nice but slow", in which it would just extract the concepts `phone`, `nice`, and `slow` (which in turn would be unlikely to result in a negative polarity on account of `nice` and `slow` bearing antithetic polarity values that nullify each other).

To this end, we further develop and apply dynamic sentic patterns [292], linguistic patterns for concept-level sentiment analysis, which allow sentiments to flow from concept to concept based on the dependency relation of the input sentence and, hence, to generate a binary (positive or negative) polarity value reflecting the

---

Part of this chapter is reprinted from Knowledge-Based Systems, 69, 45–63, Poria, Cambria, Winterstein, Huang, "Sentic patterns: Dependency based rules for conceptlevel sentiment analysis" 2014, with permission from Elsevier.

S. Poria et al., *Multimodal Sentiment Analysis*, Socio-Affective Computing 8,
https://doi.org/10.1007/978-3-319-95020-4_6

feeling of the speaker. It should be noted that, in some cases, the emotion attributed to a speaker can differ from his/her opinion. For example, (1) conveys a negative sentiment, even though the speaker conveys that he/she is satisfied. There is a gap between the informational and emotional contents of the utterance and we are interested in the latter.

(1) I am barely satisfied.

Similarly, a speaker can convey an objectively negative fact by presenting it in a positive way, as in (2).

(2) It is fortunate that Paul died a horrible death.

Irrespective of Paul's fate, the (possibly psychotic) speaker presents it as a good thing. Hence, the inferred polarity is positive. Nevertheless, in most product or service reviews, the sentiment attributed to the speaker coincides with the opinion expressed. For example, if a sentence attributes a positive property to an object (e.g., "The battery is very good"), the sentiment of the speaker is considered corresponding to his/her evaluation.

In order to dynamically compute polarity, sentic patterns leverage on a novel model for affective analogical reasoning [39] based on an artificial neural network (ANN) and on the syntactic dependency relations found in the input sentence. It is therefore an explicit approach that relies on linguistic considerations rather than on less interpretable models, such as those produced by most machine learning approaches. The upshot of this approach is that, besides being interpretable, it can take into account complex linguistic structures in a straightforward yet dynamic manner and can be easily modified and adapted.

The general template we propose for sentence-level polarity detection is illustrated in Sect. 6.2, notably by describing how polarity gets inverted (Sect. 6.2.2) and the way the calculus of polarity takes advantage of the discursive structure of the sentence (Sect. 6.2.3). The rules associated with specific dependency types are given in Sect. 6.2.7. A concrete example is given in Sect. 6.2.9.

6.2 General Rules

6.2.1 Global Scheme

The polarity score of a sentence is a function of the polarity scores associated to its sub-constituents. In order to calculate these polarities, sentic patterns dynamically consider each of the sentence's tokens by following their linear order and look at the dependency relations they have with other elements. A dependency relation is a binary relation characterized by the following features:

- The **type** of the relation that specifies the nature of the (syntactic) link between the two elements in the relation.
- The **head** of the relation: this is the element which is the pivot of the relation. Core syntactic and semantics properties (e.g., agreement) are inherited from the head.
- The **dependent** is the element that depends on the head and which usually inherits some of its characteristics (e.g., number, gender in case of agreement).

Most of the time, the active token is considered in a relation if it acts as the *head* of the relation, although some rules are an exception. Once the active token has been identified as the trigger for a rule, there are several ways to compute its contribution, depending on how the token is found in SenticNet. The preferred way is to consider the contribution not of the token alone, but in combination with the other element in the dependency relation.

This crucially exploits the fact that SenticNet is not just a polarity dictionary, but it also encodes the polarity of complex concepts. For example, in (3), the contribution of the verb `watch` will preferably be computed by considering the complex concept `watch movie` rather than the isolated concepts `watch` and `movie`.

(3) I watched a movie.

If SenticNet has no entry for the multi-word concept formed by the active token and the element related to it, then the way individual contributions are taken into account depends on the type of the dependency relation. The specifics of each dependency type are given in Sect. 6.2.7.

Since SenticNet sometimes encodes sentiment scores for a token and a specific categorization frame, dynamic sentic patterns also check whether there is an entry for a frame corresponding to the active token and the part of speech of the other term in the dependency relation.

6.2.2 *Polarity Inversion*

Once the contribution of a token has been computed, sentic patterns check whether the token is in the scope of any polarity switching operator. The primary switching operator is negation: the use of negation on a positive token (4-a) yields a negative polarity (4-b).

(4) a. I liked the movie.
    b. I did not like the movie.

However, double negation can keep the polarity of the sentence intact by flipping the polarity twice. For example, (5-a) is positive and (5-b) inverts its polarity.

However, (5-c) keeps the polarity of (5-a) identical because in (5-c) *dislike* conveys negative polarity and, hence, nullifies the negation word *not*.

(5) a. I like it.
b. I do not like it.
c. I do not dislike it.

Besides negation, other polarity switching operators include:

- exclusives such as *only, just, merely...* [74]
- adverbs that type their argument as being low, such as *barely, hardly, least...*

(6) Paul is the least capable actor of his time.

- upper-bounding expressions like *at best, at most, less than...*
- specific constructions such as the use of past tense along with a comparative form of an adjective as in (7) or counter-factuals expressed by expressions like *would/could have been*

(7) a. My old phone was better. $\rightsquigarrow$ Negative
b. My old phone was slower. $\rightsquigarrow$ Positive

Whenever a token happens to be in the scope of such an element, its polarity score is inverted. Finally, inversion also happens when some specific scopeless expressions occur in a sentence, such as *except me*.

A shortcoming of our treatment of negation is that it does not take into account the different effects of negation on various layers of meaning. It is a well known fact in linguistics that some items convey complex meanings on different layers. Presupposition is probably the most studied phenomenon of this kind: both versions of (8) convey that John killed his wife, even though the second version is the negation of the first one [21, 183].

(8) a. John regrets killing his wife.
b. John does not regret killing his wife.

In the domain of sentiment related expressions, the class of *expressives* has comparable behavior, even though these elements have been analyzed as conventional implicatures rather than presuppositions [306]. For example, a verb like *waste* can be analyzed as conveying two distinct pieces of meaning: an event of money spending and a negative evaluation regarding this spending. In some cases, this negative component is not affected by negation: (9) convey that the phone is not worth the money, even though the verb *waste* is embedded under a negation.

(9) a. I will not waste my money on this phone.
b. I do not want to waste my money on this phone.
c. I did not waste my money on this phone.

Therefore, the current treatment of negation needs to be supplemented by classification of expressions indicating whether their negative (or positive) behavior has to be analyzed as a *main content*, affected by negation and other operators, or as a *projective content*, i.e., content that 'survives' or is non-canonically affected by operators that usually affect truth-conditional content. It might prove difficult to be exhaustive in our description since projection is not a purely semantic problem but is also affected by pragmatic contextual factors [358]. Nevertheless, it is conceivable to rely on a list of elements which convey sentiment on a clearly non-main level and to tune the algorithm to deal with them.

6.2.3 Coordinated and Discourse Structures

Coordination is an informationally rich structure for which sentic patterns have rules that do not specify which elements should be looked for in SenticNet, rather they indicate how the contributions of different elements should be articulated.

In some cases, a sentence is composed of more than one elementary discourse unit (in the sense of Asher and Lascarides [7]). In such cases, each unit is processed independently and the discourse structure is exploited in order to compute the overall polarity of the sentence, especially if an overt discourse cue is present.

At the moment, we only consider structures that use an overt coordination cue and limit the analysis to adversative markers like *but* and to the conjunctions *and* and *or*.

6.2.4 But and Adversatives

Adversative items like *but*, *even though*, *however*, *although*, etc. have long been described as connecting two elements of opposite polarities. They are often considered as connecting two full-fledged discourse units in the majority of cases even when the conjuncts involve a form of ellipsis [318, 377].

It has also long been observed that, in an adversative structure, the second argument "wins" over the first one [5, 394]. For example in (10-a) the overall attitude of the speaker goes against buying the car, whereas just inverting the order of the conjuncts yields the opposite effect (10-b) while keeping the informational content identical.

(10) a. This car is nice but expensive.
b. This car is expensive but nice.

Therefore, when faced with an adversative coordination, dynamic sentic patterns primarily consider the polarity of the right member of the construction for the calculation of the polarity of the overall sentence.

In some cases, the polarity of one of the conjuncts is reported by the algorithm as undefined. This can happen for two reasons: either the conjunct is indeed neutral, or the algorithm could not determine its polarity because of incomplete dictionary. We assume here that if one of the conjuncts bears some polarity, then most probably—though not always—the other conjunct also bears some polarity even if our algorithm could not determine it; thus, we interpret undefined result as something that we did not understand rather than neutral. We also assume that usually—though not always—people use adversative conjunctions to express opposite polarity judgments. Thus, if we could determine the polarity of one of the conjuncts but not the other, then we assume, by best guess, the two conjuncts to have opposite polarities. For example, in (11).

(11) The car is nice but <cough cough cough>

we did not hear what the speaker wanted to say about the car, but we can rather safely assume that she meant something negative.

Still in rare cases, the polarity of the two conjuncts in an adversative construction can coincide—for example, in intensifying use:

(12) I expected the car to be expensive, but it was even more expensive than I could have imagined.

All possible combinations of conjuncts are summarized in Table 6.1.

Specific heuristics triggered by tense are added to this global scheme. Whenever the two conjuncts share their topic and the second conjunct is temporally anterior to the first one, the overall polarity will be that of the first conjunct. Thus, in (13) since both conjuncts are about the director and the first one is posterior, the first one drives the polarity calculus.

(13) This director is making awful movies now, but he used to be good.

Another specific rule is implemented to deal with structures combining *not only* and *but also*, as in (14).

(14) The movie is not only boring but also offensive.

Table 6.1 Adversative sentic patterns

Left conjunct	Right conjunct	Total sentence
Pos.	Neg.	Neg.
Neg.	Pos.	Pos.
Pos.	Undefined	Neg.
Neg.	Undefined	Pos.
Undefined	Pos.	Pos.
Undefined	Neg.	Neg.

In such cases, *but* cannot be considered an opposition marker. Rather, both its conjuncts argue for the same goal. Therefore, when this structure is detected, the rule applied is the same as for conjunctions using *and* (cf. infra).

6.2.5 *And*

The conjunction *and* has been described as usually connecting arguments that have the same polarity and are partly independent [163]. Therefore, when a coordination with *and* is encountered, the overall polarity score of the coordination corresponds to the sum of both conjuncts. If only one happens to have a polarity score, this score is used with the addition of a small bonus to represent the fact that *and* connects independent arguments (i.e., the idea that speakers using *and* stack up arguments for their conclusions). In case of conflicts, the polarity of the second conjunct is used.

6.2.6 *Or*

A disjunction marked by *or* is treated in the same way as the *and* disjunction, i.e., by assuming that in the case where one of the conjuncts is underspecified, its polarity is determined by the other. However, there is no added bonus to the polarity score, since the semantics of disjunction do not imply independent arguments.

6.2.7 *Dependency Rules*

In this section, we go over all of the rules that have been implemented to deal with specific dependency patterns. The main goal of these rules is to drive the way concepts are searched in SenticNet. One can roughly distinguish between two classes of dependencies:

- Relations of *complementation* where the dependent is an essential argument of the head.
- Relations of *modification* where the dependent is not sub-categorized by the head and acts as an adjunct.

We begin by focusing on essential arguments of verbs (Sect. 6.2.7.1), then move to modifiers (Sect. 6.2.7.2) and describe the rest of the rules in Sect. 6.2.7.3.

The default behavior of most rules is to build a multi-word concept formed by concatenating the concepts denoted by the head and the dependent of the relation (as exemplified in (3)). This multi-word concept is then searched in SenticNet. If it is not found, the behaviors of the rule differ.

Therefore, in our descriptions of the rules, we systematically indicate:

- what triggers the rule;
- the behavior of the rule, i.e., the way it constructs complex concepts from the parts of the dependency relation under analysis.

To simplify the notation, we will use the following notation:

- R denotes the relation type;
- h the head of the relation;
- d the dependent of the relation.

Therefore, writing $R(h, d)$ means that the head h has a dependency relation of type R with the dependent d. We use typewriter font to refer to the concept denoted by a token, e.g., `movie` is the concept denoted by both tokens *movie* and *movies*. The concepts are the elements to be searched in SenticNet.

6.2.7.1 Relations of Complementation

We consider six relations of complementation, all centered on the verb as the head of the relation. One rule deals with the subject of the verb, the other three cover the different types of object a verb can take: noun phrases, adjective or full clauses.

Subject Nouns

Trigger: When the active token is found to be the syntactic subject of a verb.
Behavior: If the multi-word concept `(h,d)` is found in SenticNet, then it is used to calculate the polarity of the relation, otherwise the following strategies are followed:

- If the sentence is in passive voice and h and d are both negative, then the subject noun relation between h and d yields positive sentiment. If the sentence is not in passive voice, then the sentiment of the relation is negative.
- If h is negative and d is positive and the speaker is a first person, then the expressed sentiment is positive, otherwise sentic patterns predict a negative sentiment.
- If h is positive and d is negative, then the expressed sentiment is detected as negative by the sentic patterns.
- If h and d are both positive, then the relation results in a positive sentiment.

Example 1: In (15), *movie* is in a subject noun relation with *boring*.

(15) The movie is boring.

If the concept `(movie, boring)` is in SenticNet, its polarity is used. Otherwise, sentic patterns dynamically perform a detailed analysis of the relation to obtain the polarity. In this case, sentiment of `h` is treated as the sentiment of the relation.

Example 2: In (16), *relieve* is in a subject noun relation with *trouble*. Here, the polarity of *trouble* is negative and the polarity of *relieve* is positive. According to our rule, sentiment is carried by the *relieved*. So, here the sentence expresses a positive sentiment.

(16) His troubles were relieved.

Example 3: In (17), *success* is in subject noun relation with *pissed*. The polarity of *success* is positive while *pissed* has negative polarity. The final polarity of the sentence is negative according to our rule.

(17) My success has pissed him off.

Example 4: In (18), *gift* is in subject noun relation with *bad*. The polarity of *gift* is positive and *bad* is negative. Therefore, sentic patterns dynamically extract the polarity of the sentence as negative.

(18) Her gift was bad.

Direct Nominal Objects

This complex rule deals with direct nominal objects of a verb. Its complexity is due to the fact that the rule attempts to determine the modifiers of the noun in order to compute the polarity.

Trigger: When the active token is head verb of a direct object dependency relation.

Behavior: Rather than searching directly for the binary concept `(h,d)` formed by the head and the dependent, the rule first tries to find richer concepts by including modifiers of the nominal object. Specifically, the rule searches for relative clauses and prepositional phrases attached to the noun and if these are found, it searches for multi-word concepts built with these elements. Thus, if the dependent d is head of a relation of modification $R'(d, x)$, then sentic patterns will consider the ternary concept `(h,d,x)`. If this procedure fails and the binary concept `(h,d)` is not found either, the sign of the polarity is preferably driven by the head of the relation.

Example 1: In (19), sentic patterns first look for `(see,movie,in 3D)` in the SenticNet and, if this is not found, they search for `(see,movie)` and then `(see, in 3D)`.

(19) Paul saw the movie in 3D.

`(movie,in 3D)` is not considered at this stage since it will be analyzed later under the standard rule for prepositional attachment. If the searching process fails, the polarity will be the one of `see` and eventually `movie`.

Example 2: In (20), first the concept `(make, pissed)` is searched in SenticNet and since it is not found, the dynamic sentic patterns look for the polarity of *make* and *pissed* separately. As *make* does not exist in the SenticNet, the polarity of *pissed* is considered as the polarity of the sentence (which is negative).

(20) You made me pissed off.

Example 3: In (21), the polarity of *love* is positive and the polarity of *movie* is negative as it is modified by a negative modifier *boring*. Sentic patterns set the polarity of this sentence as negative as the speaker says it is a boring movie though the subject *John* loves it.

(21) John loves this boring movie.

This rule has an exception when the subject is first person, i.e., the subject of the sentence and the speaker are the same.

Example 4: In (22), *hurt* has negative polarity and the polarity of *cat* is positive as it has a positive modifier *cute*. Thus, according to sentic patterns, the polarity of the sentence is negative.

(22) You have hurt the cute cat.

Complement Clause

This rule is fired when a sentence contains a finite clause which is subordinate to another clause: "That" and "Whether" are complement clauses.

Trigger: When a complement clause is found in a sentence.
Behavior: The sentence is split into two parts based on the complement clause:

- The sentiment expressed by the first part is considered as the final overall sentiment.
- If the first part does not convey any sentiment, then the sentiment of the second part is taken as the final sentiment.
- If the first part does not express any sentiment but a negation is present, then the sentiment of the second part is flipped.

Example 1: In (23), the sentiment expressed by the part of the sentence before "that" is positive, so the overall sentiment of the sentence is considered positive.

(23) I love that you did not win the match.

Example 2: In (24), the portion of the sentence before "whether" has no sentiment, but it contains a negation which alters the polarity of the second part. Thus, the overall polarity of the sentence becomes negative.

(24) I do not know whether he is good.

Adverbial Clause

Trigger: When a sentence contains an adverbial clause (i.e., "while").
Behavior: The role of "while" in a sentence is similar to the one of "but". Then, the sentic patterns first split the sentence into two parts by recognizing the subject and the use of comma in the sentence. Then, the overall sentiment of the sentence is conveyed by the second part.
Example: In (25), sentic patterns first identify the two parts of the sentence by recognizing the comma and the subject after the comma. The polarity of the first part (i.e., *i'm sure the quality of the product is fine*) is positive but the polarity of the second part (*the color is very different*) is neutral. Sentic patterns therefore detect the polarity of the sentence as negative.

(25) While I'm sure the quality of the product is fine, the color is very different.

Adjective and Clausal Complements

These rules deal with verbs having as complements either an adjective or a closed clause (i.e., a clause, usually finite, with its own subject).

Trigger: When the active token is head verb of one of the complement relations.
Behavior: First, sentic patterns look for the binary concept `(h,d)`. If it is found, the relation inherits its polarity properties. If it is not found:

- If both elements `h` and `d` are independently found in SenticNet, then the sentiment of `d` is chosen as the sentiment of the relation.
- If the dependent `d` alone is found in SenticNet, its polarity is attributed to the relation.

Example: In (26), *smells* is the head of a dependency relation with *bad* as the dependent.

(26) This meal smells bad.

The relation inherits the polarity of `bad`.

Open Clausal Complements

Open clausal complements are clausal complements of a verb that do not have their own subject, i.e. they usually share their subjects with the ones of the matrix clause.

Trigger: When the active token is the head predicate of the relation.[1]

Behavior: As for the case of direct objects, sentic patterns try to determine the structure of the dependent of the head verb. Here the dependent is itself a verb, therefore, sentic patterns attempt to establish whether a relation $R'(d, x)$ exists, where x is a direct object or a clausal complement of d. Sentic patterns are therefore dealing with three elements: the head/matrix verb (or predicate) h, the dependent predicate d, and the (optional) complement of the dependent predicate x. Once these have been identified, sentic patterns first test the existence of the ternary concept `(h,d,x)`. If this is found in SenticNet, the relation inherits its properties. If it is not found, sentic patterns check for the presence of individual elements in SenticNet.

- If `(d,x)` is found as well as `h` or if all three elements `h`, `d` and `x` are independently found in SenticNet, then the final sentiment score will be the one of `(d,x)` or it will be calculated from `d` and `x` by following the appropriate rule. The head verb affects the sign of this score. The rules for computing the sign are summarized in Table 6.2, where the final sign of the score is expressed as a function of the signs of the individual scores of each of the three relevant elements.
- If the dependent verb `d` is not found in SenticNet but the head verb `h` and the dependent's complement `x` can be found, then they are used to produce a score with a sign again corresponding to the rules stated in Table 6.2.

Table 6.2 Polarity algebra for open clausal complements

Matrix predicate (h)	Dependent predicate (d)	Dep. comp. (x)	Overall polarity	Example
Pos	Pos	Pos	Pos	(27-a)
Pos	Pos	Neg	Neg	(27-b)
Pos	Neg	Pos	Neg	(27-c)
Pos	Neg	Neg	Pos	(27-d)
Neg	Pos	Pos	Neg	(27-e)
Neg	Pos	Neg	Neg	(27-f)
Neg	Neg	Pos	Neg	(27-g)
Neg	Neg	Neg	Neg	(27-h)
Pos	Neutral	Pos	Pos	(27-i)
Pos	Neutral	Neg	Neg	(27-j)
Neg	Neutral	Pos	Neg	(27-k)
Neg	Neutral	Neg	Neg	(27-l)

[1]Usually the token is a verb, although when the tensed verb is a copula, the head of the relation is rather the complement of the copula.

Example: In order to illustrate every case presented in Table 6.2, we use the paradigm in (27). For each example, the final sign of the polarity is calculated according to Table 6.2. The examples assume the following:

- *h*, the matrix predicate, is either:
  - *perfect*, which has a positive polarity
  - *useless*, which has a negative polarity
- *d*, the dependent verb, is either:
  - *gain*, which has a positive polarity
  - *lose*, which has a negative polarity
  - *talk*, which is not found isolated in SenticNet, i.e., is considered neutral here
- *x*, the complement of the dependent verb, is either:
  - *money*, which has a positive polarity
  - *weight*, which has a negative polarity[2]

It must be remembered that for such examples we assume that the sentiment expressed by the speaker corresponds to his/her opinion on whatever *this* refers to in the sentence: if the speaker is positive about the thing he/she is talking about, we consider that he/she is expressing positive sentiments overall.

(27) a. This is perfect to gain money.
b. This is perfect to gain weight.
c. This is perfect to lose money.
d. This is perfect to lose weight.
e. This is useless to gain money.
f. This is useless to gain weight.
g. This is useless to lose money.
h. This is useless to lose weight.
i. This is perfect to talk about money.
j. This is perfect to talk about weight.
k. This is useless to talk about money.
l. This is useless to talk about weight.

[2]The negative score associated with *weight* does not reflect a deliberate opinion on the meaning of term. This score is extracted from SenticNet and has been automatically computed as explained in [45]. Thus, even though the term might not appear negative at first glance, its sentiment profile is nevertheless biased towards the negative.

6.2.7.2 Modifiers

Modifiers, by definition, affect the interpretation of the head they modify. This explains why in most of the following rules the dependent is the guiding element for the computation of polarity.

Adjectival, Adverbial and Participial Modification

The rules for items modified by adjectives, adverbs or participles all share the same format.

Trigger: When the active token is modified by an adjective, an adverb or a participle.

Behavior: First, the multi-word concept `(h,d)` is searched in SenticNet. If it is not found, then the polarity is preferably driven by the modifier `d`, if it is found in SenticNet, otherwise `h`.

Example: In (28), both sentences involve elements of opposite polarities. The rule ensures that the polarity of the modifiers is the one that is used, instead of the one of the head of the relation: e.g., in (28-b) *beautifully* takes precedence over *depressed*.

(28) a. Paul is a bad loser.
b. Mary is beautifully depressed.

Unlike other NLP tasks such as emotion recognition, the main aim of sentiment analysis is to dynamically infer the polarity expressed by the speaker (i.e., the person who writes the review of a hotel, product, or service). Hence, a sentence such as (28-b) would be positive as it reflects the positive sentiment of the speaker.

Relative Clauses

Trigger: When the active token is modified by a relative clause, restrictive or not. The dependent is usually the verb of the relative clause.

Behavior: If the binary concept `(h,d)` is found in SenticNet, then it assigns polarity to the relation, otherwise the polarity is assigned (in order of preference):

- by the value of the dependent verb `d` if it is found in SenticNet.
- by the value of the active token `h` if it is found.

Example: In (29), *movie* is in relation with *love* which acts as a modifier in the relative clause.

(29) I saw the movie you love.

Assuming `(love, movie)` is not in SenticNet while `love` is, then the latter will contribute to the polarity score of the relation. If none of these is in SenticNet, then the dependency will receive the score associated with `movie`. In the case of (29), the polarity will be inherited at the top level because the main verb *see* is neutral. However, the overall polarity of a sentence like (30) is positive since, in case the subject is a first person pronoun, the sentence directly inherits the polarity of the main verb, here *like* (see Sect. 6.2.7.3 for more details).

(30) I liked the movie you love.

Similarly, (31) will obtain an overall negative sentiment because the main verb is negative.

(31) I disliked the movie you love.

Prepositional Phrases

Although prepositional phrases (PPs) do not always act as modifiers, we insert them in this section since the distinction is not significant for their treatment. Another reason is due to the fact that the Stanford dependency parser on which we rely does not differentiate between modifier and non-modifier PPs.

Trigger: The rule is activated when the active token is recognized as typing a prepositional dependency relation. In this case, the head of the relation is the element to which the PP attaches, and the dependent is the head of the phrase embedded in the PP. This means that the active element is not one of the two arguments of the relation but participates in the definition of its type.

Behavior: Instead of looking for the multi-word concept formed by the head h and the dependent d of the relation, sentic patterns use the preposition $prep$ (corresponding to the active token) to build a ternary concept `(h, prep, d)`. If this is not found, then they search for the binary concept `(prep, d)` formed by the preposition and the dependent and use the score of the dependent `d` as a last tentative. This behavior is overridden if the PP is found to be a modifier of a noun phrase (NP) that acts as the direct object.

Example 1: In (32), the parser yields a dependency relation using *with* between the verb *hit* and the noun *hammer* (=the head of the phrase embedded in the PP).

(32) Bob hit Mary with a hammer.

Therefore, sentic patterns first look for the multi-word concept `(hit, with, hammer)` and, if this is not found, they look for `(with, hammer)` and finally `hammer` itself.

Example 2: In (33), the PP headed by *in* is a modifier of the verb *complete*, which is positive in SenticNet. *Terrible way* is however negative and, because it directly modifies the verb, the overall polarity is given by this element.

(33) Paul completed his work in a terrible way.

Example 3: In (34), the PP introduced by *in* is attached to the direct object of the predicate *is a failure*.

(34) This actor is the only failure in an otherwise brilliant cast.

Here, sentic patterns will ignore the contribution of the PP since the main sentiment is carried by the combination of the verb and its object, which is negative.

Adverbial Clause Modifier

This kind of dependency concerns full clauses that act as modifiers of a verb. Standard examples involve temporal clauses and conditional structures.

Trigger: The rule is activated when the active token is a verb modified by an adverbial clause. The dependent is the head of the modifying clause.

Behavior: If the binary concept `(h,d)` is found in SenticNet, then it is used for calculating the score. Otherwise, the rule assigns polarity by considering first the dependent `d`, then the head `h`.

Example: In (35), *playing* modifies *slows*. If the multi-word concept `(slow, play)` is not in SenticNet, then first `play` then `slow` will be considered.

(35) The machine slows down when the best games are playing.

Untyped Dependency

Sometimes the dependency parser detects two elements that keep a dependency relation but it is unable to type it properly. In this case, if the multi-word concept `(h,d)` is not found, the polarity is computed by considering the dependent `d` alone.

6.2.7.3 Other Rules

First Person Heuristics

On top of the rules presented so far, we implemented a specific heuristic when the subject of a sentence is the first person pronoun. In this case, the sentiment is essentially carried by the head verb of the relation. We can analyze the contrast in (36):

(36) a. Paul likes bad movies.
b. I like bad movies.

Whereas (36-a) is a criticism of Paul and his tastes, (36-b) is speaker-oriented as he/she expresses his/her (maybe peculiar) tastes. What matters is that the speaker of (36-b) is being positive and uses the verb *like*. This overrides the calculus that would yield a negative orientation as in (36-a) by considering the combination of `like` and `bad movies`.

Similarly, in (37) the use of the first person overrides the effect produced by the relative clause *which you like*. The overall sentiment is entirely driven by the use of the verb *hate* which is negative.

(37) I hate the movie which you like.

Rule for the Preposition "against"

In English, "against" is a preposition which carries a sentiment. Usually it is used as a negative sentiment expressing word. But, *against* can also be used in a sentence to express positive sentiment. Here, we give a few examples to explain the role of "against" in determining the sentiment of a sentence. In (38), *activity* has negative sentiment as it is modified by a negative modifier, i.e., *criminal*. Here, *against*, attached to the target *activity*, actually flips the polarity of *activity* and the overall sentiment of the sentence becomes positive.

(38) I am against all criminal activities.

In (39), *against* attaches to the target *love* which has positive polarity. Then, the overall sentiment of the sentence becomes negative.

(39) He is against me and your love.

If *against* attaches to a word with no polarity then the sentence sentiment turns negative.

6.2.8 Activation of Rules

The algorithm dynamically operates over the dependency parse tree of the sentence. Starting from the first (leftmost) relation in the tree, the rules corresponding to relations are activated: for a relation $R(A, B)$, the rules of the form R_i are activated to assign polarity (not necessarily the same) to the relation itself and to the words A and B. The rules for relations that involve either A or B are scheduled to be activated next; the main idea of the algorithm is taking into account the polarity already assigned to the relations and words previously processed. However, a rule may alter the order of activation of other rules if it needs additional information before it can proceed. For example, while computing the polarity of a relation $R(A, B)$, if A and B have any modifier, negation and subject-noun relation, then those relations are computed immediately. The reason is that such relations may alter the polarity of A and B.

If there is no rule for a given relation $R(A, B)$, then it is left unprocessed and the new relations are scheduled for processing using the method described above. When there are no relations scheduled for processing, the process restarts from the leftmost relation not yet processed for which a rule exists.

The output of the algorithm is the polarity of the relation processed last. It dynamically accumulates the information of all relations in the sentence, because each rule takes into account the result of the previous ones, so that the information flows from the leftmost relation towards the rule executed last, which often corresponds to one of the rightmost relations.

Below, for (40). we describe the sentiment flow across the dependency arcs based on the sentic patterns.

(40) My failure makes him happy.

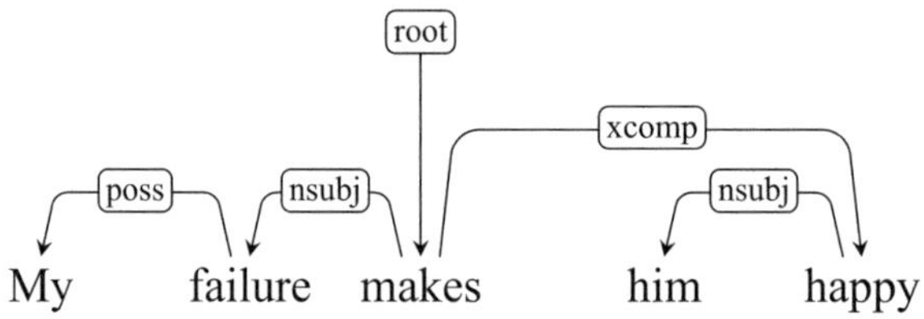

- First the relation between *my* and *failure* is considered. This is a *possession modifier* relation which does not satisfy any rule, so nothing has to be done.
- Then, the algorithm computes the polarity of the subject-noun relation between *make* and *failure*. The sentiment of this relation is negative according to the sentic patterns. The rule also assigns negative polarity to *make* which actually is a neutral word. This polarity is a contextual polarity to be used to compute the polarity of subsequent relations.
- Next, the polarity of the relation between *make* and *happy* is computed. This computation needs also the polarity of the relation computed in the previous step. Before computing the polarity of this relation, the subject-noun relation between *him* and *happy* is computed and a positive polarity is obtained. This polarity value does not alter the polarity of *happy*, which is positive according to SenticNet. *Make* has a negative polarity according to the previous step. Then, there is a clausal complement relation between *make* and *happy*. Based on the clausal complement rule, sentic patterns assigns negative polarity to this relation. After this computation there is no more relation left which satisfies the rules, so the sentence is assigned negative polarity by the algorithm.

(41) is another example to show the activation of rules and the flow of sentiments across the dependency arcs.

(41) You hurt the beautiful cat.

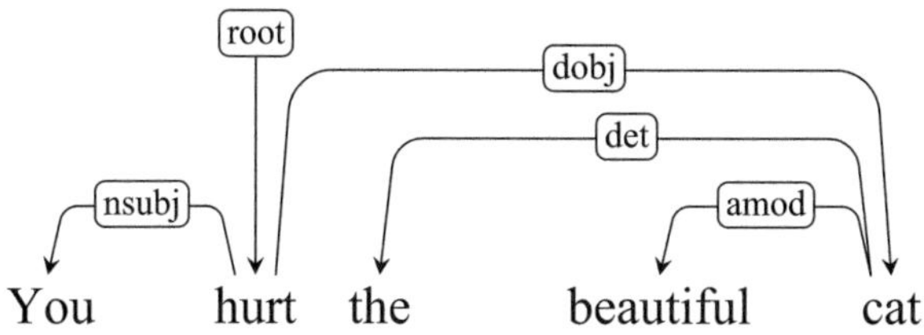

- First the algorithm encounters a subject-noun relation between *you* and *hurt*. As the polarity of *hurt* is negative, the algorithm assigns negative sentiment to the relation and *hurt* also maintains its negative polarity.
- Next, the algorithm finds *hurt* in a direct object relation with *cat*. To obtain the polarity of this relation, the algorithm first obtains the polarity of *cat* and the polarity of *hurt*, which was computed in the previous step. *Cat* does not exist in SenticNet but *cat* is modified by a positive word *beautiful*. So, *cat* is assigned positive polarity by sentic patterns. To compute the polarity of the direct object relation between *hurt* and *cat*, the algorithm has now all the necessary information. Based on the sentic patterns, it assigns negative polarity to this relation.
- The relation between *the* and *cat* does not satisfy any rule in sentic patterns. Nothing is done and there is no other relation to be processed. The final polarity of the sentence becomes negative (Fig. 6.1).

6.2.9 Walking Through an Example

Here we describe how the global sentiment for a complex example is computed. This is made in order to show how the sentiment flows in the treatment of a sentence. We will base our presentation on the (artificial) case of (42). Figure 6.2 shows the parse tree for the sentence (42).

(42) The producer did not understand the plot of the movie inspired by the book and preferred to use bad actors.

The relevant dependency relations here are (with the concept arguments given between parenthesis):

1. A general coordination with *and (understand, preferred)*
2. Two subject relations *(understand, producer)* and *(preferred, producer)*

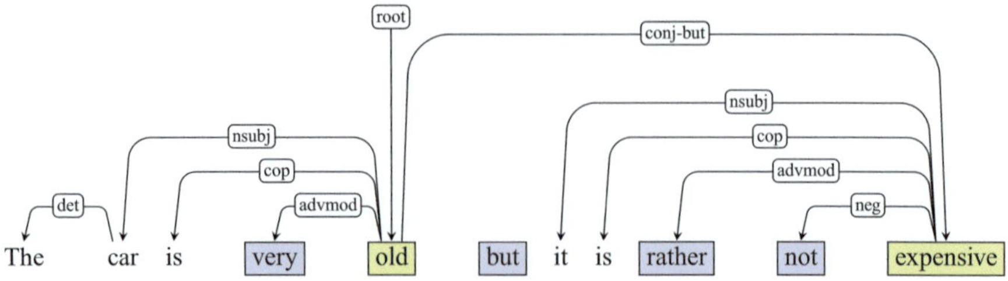

(a) Dependency tree of a sentence.

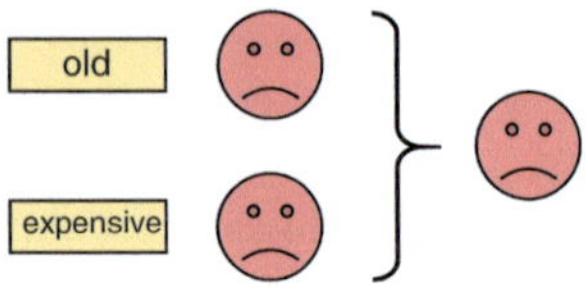

(b) The old way: averaging over a bag of sentiment words.
The overall polarity of a sentence is given by the algebraic sum of the polarity values associated with each affect word divided by the total number of words.

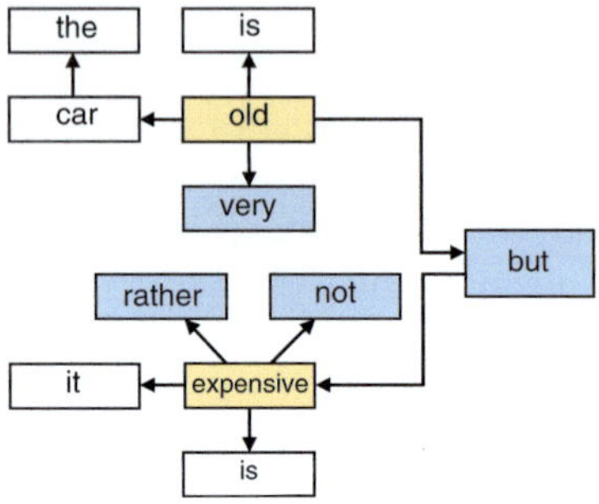

(c) The dependency tree of a sentence resembles an electronic circuit: words shown in blue can be thought as a sort of "boolean operations" acting on other words.

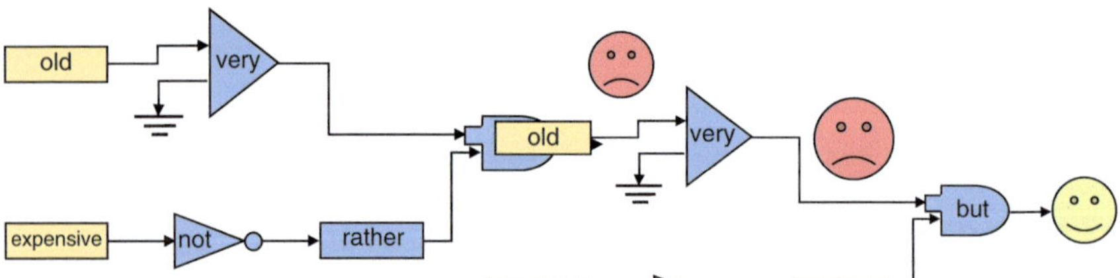

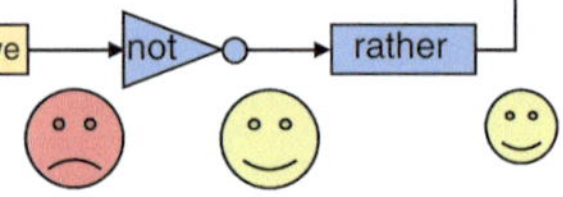

(d) The electronic circuit metaphor: sentiment words are "sources" while other words are "elements", e.g., *very* is an amplifier, *not* is a logical complement, *rather* is a resistor, *but* is an OR-like element that gives preference to one of its

(e) The final sentiment data flow of the "signal" in the "circuit".

Fig. 6.1 The main idea behind sentic patterns: the structure of a sentence is like an electronic circuit where logical operators channel sentiment data-flows to output an overall polarity. (**a**) Dependency tree of a sentence. (**b**) The old way: averaging over a bag of sentiment words. The overall polarity of a sentence is given by the algebraic sum of the polarity values associated with each affect word divided by the total number of words. (**c**) The dependency tree of a sentence resembles an electronic circuit: words shown in blue can be thought as a sort of "boolean operations" acting on other words. (**d**) The electronic circuit metaphor: sentiment words are "sources" while other words are "elements", e.g., *very* is an amplifier, *not* is a logical complement, *rather* is a resistor, *but* is an OR-like element that gives preference to one of its inputs. (**e**) The final sentiment data flow of the "signal" in the "circuit"

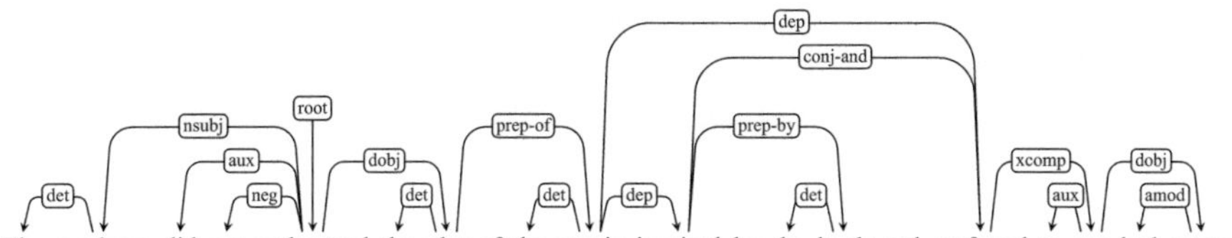

Fig. 6.2 Dependency tree for the sentence *The producer did not understand the plot of the movie inspired by the book and preferred to use bad actors*

3. A direct object relation *(understand, plot)*
4. A prepositional attachment typed by *of (plot, movie)*
5. A participial modification *(plot, inspired)*
6. An open clausal complement relation *(preferred, use)*
7. Another direct object relation *(use, actors)*
8. An adjective modifying a noun *(actors, bad)*

First, the discourse structure parser detects that the sentence has two discourse units conjoined by *and*. The final polarity will thus be a function of the elements π_1 =*The producer did not understand the plot of the movie based on the book* and π_2 =*[the producer] preferred to use bad actors*.

The computation of π_1 entails checking the relations in the following order:

- The subject relation *(understand, producer)* is considered to check whether the multi-word concept `(producer understand)` can be found in SenticNet. This is not the case, so nothing is done.
- The relations having the verb *understand* as their head are explored. Here there is only the direct object relation. In this relation the dependent object is modified in two ways:
  - by a prepositional phrase
  - by a participial modifier

  Thus, sentic patterns will first try to find the multi-word concept `(understand, plot, of, movie)`. Since this one is not found, `(understand, plot, inspired)` is tried, and it is not in SenticNet either. Finally, sentic patterns fall back on the concept `(understand, plot)`, which is found in SenticNet. Therefore, the polarity stack is set at the corresponding positive value.
- Since the previous polarity is in the scope of a sentential negation, the sign of the previous score is switched to assign a negative value.

Now sentic patterns analyze π_2.

- The open clausal modification rule determines the dependent of the dependent. This case means identifying *actors* as the direct object of *use*.
- Since *actors* is modified by *bad*, it will inherit its negative orientation.

- The only relevant elements for the computation of the polarity, due to the open clausal complement, are *prefer* (which is positive) and *actor* (negative because of its adjectival modification). Therefore, the final polarity score is also negative.

In the end, both the *and* conjuncts are negative, meaning that the overall polarity of the sentence is also negative with a value equal to the sum of the scores of each conjunct.

6.3 Combining Sentic Patterns with Machine Learning for Text-Based Sentiment Analysis

Despite being much more efficient than BoW and BoC models, Sentic Patterns is still limited by the richness of the knowledge base and the set of dependency-based rules. In order to be able to make a good guess even when no sentic pattern is matched or SenticNet entry found, we resort to Machine Learning. In particular, we use three well-known sentiment analysis datasets (Sect. 6.3.2), a set of features per sentence (Sect. 6.3.1) and an artificial neural network (ANN) classifier (Sect. 6.3.3) to label stretches of text as positive or negative. Each of these are briefly described, next.

6.3.1 Feature Extraction

Various features were extracted from the sentences for building the machine learning classifier. Below, each of these features are attributed in detail. The performance of these features are given in the Sect. 6.3.3.

6.3.1.1 Deep Textual Features

Being motivated by the state of the art [283], CNN was used to automatically extract features from the text. The trained CNN features were then fed into a SVM for classification. So, in particular we used CNN as trainable feature extractor and SVM as a classifier.

The intuition for building this hybrid classifier SVM-CNN is to combine the merits of each classifier and form a hybrid classifier to enhance accuracy. Recent studies [317] also show the use of CNN for feature extraction. In theory, the training process of CNN is similar to MLP as CNN is an extension of traditional MLP. MLP network is trained using a backpropagation algorithm which uses Empirical Risk Minimization. It tries to minimize the errors in training data. Once it finds the hyperplane, regardless of global or local optimum, the training process is stopped. This means that it does not try to improve the separation of the instances from the hyperplane. Wherein, SVM tries to minimize the generalization error on unseen data based on Structural Risk Minimization algorithm using a fixed probability distribution on training data. It therefore aims to maximize the distance between training instances and hyperplane, so the margin area between two separate training classes is maximized. This separating hyperplane is a global optimum solution. So, SVM is more generalized than MLP which enhances the classification accuracy.

On the other hand, CNN automatically extracts key features from the training data. It grasps contextual local features from a sentence and after several convolution operations it finally forms a global feature vector out of those local features. CNN does not need the hand-crafted features used in a traditional supervised classifier. The hand-crafted features are difficult to compute and a good guess for encoding the features is always necessary in order to get satisfactory result. CNN uses a hierarchy of local features which are important to learn context. The hand-crafted features often ignore such a hierarchy of local features. Features extracted by CNN can therefore be used instead of hand-crafted features, as they carry more useful information.

The hybrid classifier SVM-CNN therefore inherits the merits from each classifier and should produce a better result.

The idea behind convolution is to take the dot product of a vector of k weights w_k also known as kernel vector with each k-gram in the sentence $s(t)$ to obtain another sequence of features $c(t) = (c_1(t), c_2(t), \ldots, c_L(t))$.

$$c_j = w_k{}^T.\mathbf{x}_{i:i+k-1} \tag{6.1}$$

We then apply a max pooling operation over the feature map and take the maximum value $\hat{c}(t) = \max\{\mathbf{c}(t)\}$ as the feature corresponding to this particular kernel vector. Similarly, varying kernel vectors and window sizes are used to obtain multiple features [176].

For each word $x_i(t)$ in the vocabulary, an d dimensional vector representation is given in a look up table that is learned from the data [231]. The vector representation of a sentence is hence a concatenation of vectors for individual words. Similarly we can have look up tables for other features. One might want to provide features other than words if these features are suspected to be helpful. The convolution kernels are then applied to word vectors instead of individual words.

We use these features to train higher layers of the CNN, to represent bigger groups of words in sentences. We denote the feature learned at hidden neuron h in layer l as F_h^l. Multiple features may be learned in parallel in the same CNN layer. The features learned in each layer are used to train the next layer

$$F^l = \sum_{h=1}^{n_h} w_k^h * F^{l-1} \tag{6.2}$$

where * indicates convolution and w_k is a weight kernel for hidden neuron h and n_h is the total number of hidden neurons. The CNN sentence model preserves the order of words by adopting convolution kernels of gradually increasing sizes that span an increasing number of words and ultimately the entire sentence. Each word in a sentence was represented using word embedding and part-of-speech of that word. The details are as follows:

- **Word Embeddings –** We employ the publicly available word2vec vectors that were trained on 100 billion words from Google News. The vectors have dimensionality 300 trained using the continuous bag-of-words architecture [231]. Words not present in the set of pre-trained words are initialized randomly.
- **Part of Speech –** The part of speech of each word was also appended to the word's vector representation. As there are a total of 6 part of speech, so the length of part of speech vector was 6.

So, in the end a word was represented by a 306 dimensional vector.

Network Configuration

Each sentence was wrapped to a window of 50 words to reduce the number of parameters and hence over-fitting the model. The CNN we developed in our experiment had two convolution layers, a kernel size of 4,5 and 50 feature maps was used in the first convolution layer and a kernel size 2,3 and 100 feature maps in the second. It should be noted that the output of each convolution hidden layer is computed using a non-linear function (in our case we use ReLU). Each convolution layer was followed by a max-pool layer. The max-pool size of the first and second max-pool layer was 2. The penultimate max-pool layer is followed by a fully connected layer with softmax output. We used 500 neurons in the full connected layer. The output layer corresponded to two neurons for each class of sentiments.

We used the output of the fully connected layer (layer 6) of the network as our feature vector.

6.3.1.2 Other Sentence Level Textual Features

We have ultimately fed the features extracted by CNN to the SVM. Below are the features which were also used along with the deep textual features for the final classification.

- **Common-Sense Knowledge Features –** Common-sense knowledge features consist of concepts are represented by means of EmoSenticSpace (as explained in Chap. 5). In particular, concepts extracted from text through the semantic parser (as explained in Chap. 4) are encoded as 100-dimensional real-valued vectors and then aggregated into a single vector representing the sentence by coordinate-wise summation:

$$x_i = \sum_{j=1}^{N} x_{ij},$$

where x_i is the i-th coordinate of the sentence's feature vector, $i = 1, \ldots, 100$; x_{ij} is the i-th coordinate of its j-th concept's vector, and N is the number of concepts in the sentence.

- **Sentic Feature –** The polarity scores of each concept extracted from the sentence were obtained from SenticNet and summed up to produce a single scalar feature.
- **Modification Feature –** This is a single binary feature. For each sentence, we obtained its dependency tree from the dependency parser. This tree was analyzed to determine whether there is any word modified by a noun, adjective, or adverb. The modification feature is set to 1 in case of any modification relation in the sentence; 0 otherwise.
- **Negation Feature –** Similarly, the negation feature is a single binary feature determined by the presence of any negation in the sentence. It is important because the negation can invert the polarity of the sentence.

6.3.2 Datasets Used

6.3.2.1 Movie Review Dataset

We used a dataset derived from the benchmark corpus developed by Pang and Lee [261]. This corpus includes 1,000 positive and 1,000 negative movie reviews authored by expert movie reviewers, collected from rottentomatos.com, with all text converted to lowercase and lemmatized, and HTML tags removed. Originally, Pang and Lee manually labeled each review as positive or negative. Later, Socher et al. [361] annotated this dataset at sentence level. They extracted 11,855 sentences from the reviews and manually labeled them using a fine grained inventory of five sentiment labels: *strong positive*, *positive*, *neutral*, *negative*, and *strong negative*.

Since in this work we considered only binary classification, we removed from the dataset, sentences marked as neutral, and reduced the labels on the remaining sentences to positive or negative. Thus, our final movie dataset contained 9,613 sentences, of which 4,800 were labeled as positive and 4,813 as negative.

6.3.2.2 Blitzer Dataset

We also used the dataset introduced by Blitzer et al. [25], which consists of product reviews in seven different domains. For each domain there are 1,000 positive and 1,000 negative reviews. We only used reviews under the *electronics* category, and randomly extracted from them 7,210 non-neutral sentences, 3505 sentences from positive reviews and 3,505 from negative ones, and manually annotated them as positive or negative. Note that the polarity of individual sentences does not always coincide with the overall polarity of the review: for example, in a negative review we found sentences such as "This is a good product – sounds great", "Gets good battery life", "Everything you'd hope for in an iPod dock" or "It is very cheap", which we labeled as positive. Specifically, we obtained 3,800 sentences marked as positive and 3,410 as negative.

6.3.2.3 Amazon Product Review Dataset

We crawled the reviews of 453 mobile phones from http://amazon.com. Each review was split into sentences, and each sentence then manually labelled by its sentiment labels. Finally, we obtained 115,758 sentences, out of which 48,680 were negative, 2,957 sentences neutral and 64,121 positive. In our experiment we only employed positive and negative sentences. So, the final *Amazon dataset* contained 112,801 sentences annotated as either positive or negative.

6.3.3 Classification

We collected 60% of the sentences from each of the three datasets in order to create the training set for the classification. The sentences from each dataset were randomly drawn in such a way to balance the dataset with 50% negative sentences and 50% positive sentences. We obtained an overall 76.33% accuracy on the Final Dataset described in Table 6.3 using SVM and 64.10% accuracy using SVM.

We also trained the classifiers on each single dataset and tested over all the other datasets. Table 6.4 reports the comparative performance results obtained in this experiment. It can be noted from Table 6.4 that the model trained on the Amazon dataset produced the best accuracy compared to the movie review and Blitzer-derived datasets. For each of these experiments, SVM outperformed ELM on the feature set given in Sect. 6.3.1.2. However, it can be noticed that when deep textual features are were the performance of both ELM and SVM were almost same. The best performance by the ELM classifier was obtained on the movie review dataset, while the SVM classifier performed best on the Blitzer dataset. The training and test set collected from different datasets are shown in Table 6.3.

Table 6.3 Dataset to train and test CI classifiers

Dataset	Number of training sentences	Number of test sentences
Movie Review Dataset	5678	3935
Blitzer-Derived Dataset	4326	2884
Amazon Dataset	67681	45120
Final Dataset	77685	51939

Table 6.4 Performance of the classifiers when all features were used

Training dataset	Classifier	On movie review dataset	On Blitzer dataset	On Amazon dataset
Movie review	SVM	–	75.00%	72.63%
	ELM classifier	–	74.85%	71.40%
Blitzer	SVM	70.67%	–	65.10%
	ELM classifier	70.30%	–	65.28%
Amazon	SVM	74.59%	76.00%	–
	ELM classifier	74.27%	75.46%	–

Table 6.5 Feature analysis

Features used	Accuracy
All – All	76.33%
All – deep textual features	71.32%
All – common-sense knowledge	40.11%
All – sentic feature	70.84%
All – part-of-speech feature	70.41%
All – modification feature	71.53%
All – negation feature	68.97%

Hence, whenever we are unable to process a sentence through SenticNet and sentic patterns, it is possible to train a supervised classifier to make a good guess on the sentence polarity, based on the available features.

Though the supervised classifier has performed best when all features were used together, deep textual features and common-sense-knowledge based features resulted in the most significant ones. From the Table 6.5, it can be noticed that the negation is also a useful feature. The other features were not found to have a significant role in the performance of the classifier but were still found to be useful for producing optimal accuracy. On the features explained in Sect. 6.3.1.2, the best accuracy was obtained using the SVM classifier. Table 6.5 presents the accuracy of the SVM classifier.

6.4 Evaluation

The proposed approach (available as a demo at http://sentic.net/demo) was tested on three datasets: the movie review dataset described in Sect. 6.3.2.1, the Blitzer-derived dataset described in Sect. 6.3.2.2 and the Amazon dataset described in Sect. 6.3.2.3. As shown by results below, the best accuracy is achieved when applying an ensemble of knowledge-based analysis (Sect. 6.2) and machine-learning classification (Sect. 6.3), as the latter can act as reserve for the former when no match is found in SenticNet (Fig. 6.3). Figure 6.4 shows a comparative view of the experimental results.

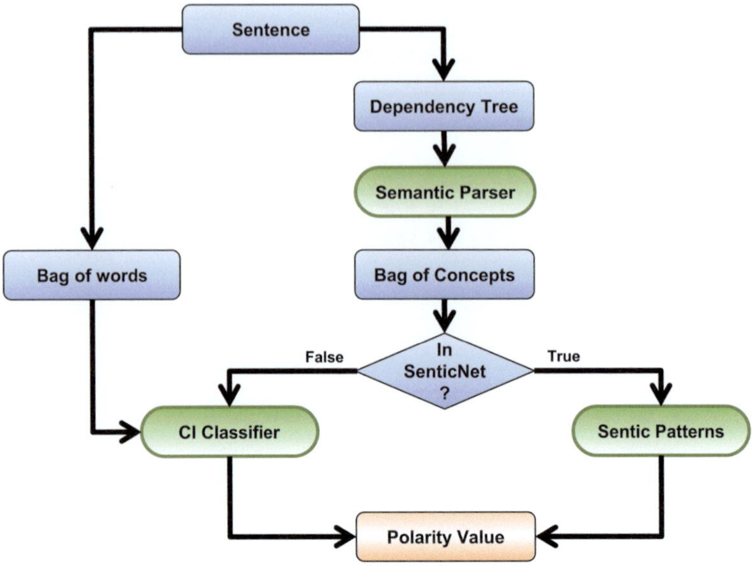

Fig. 6.3 Flowchart of the sentence-level polarity detection system. Natural language text is first deconstructed into concepts. If these are found in SenticNet, sentic patterns are applied dynamically. If none of the concepts is available in SenticNet, the supervised classifier is employed

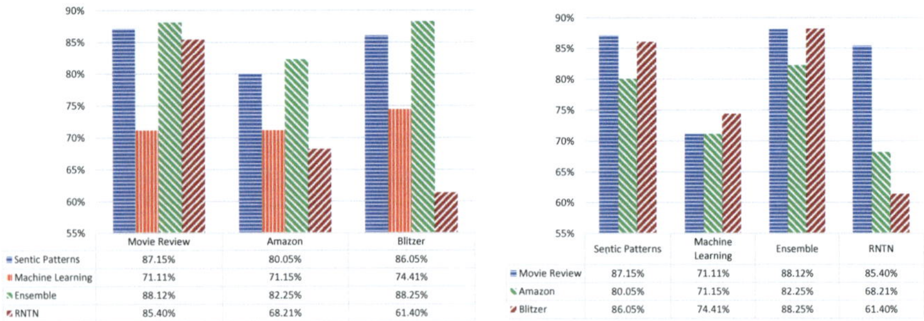

Fig. 6.4 Performance of different algorithms on the benchmark datasets

6.4.1 Results

In this section we present the experimental results obtained by combining supervised classifier with sentic patterns. As a supervised classifier we used SVM since it gave the overall best performance.

6.4.1.1 Results on the Movie Review Dataset

We evaluated our approach on the movie review dataset and obtained an accuracy of 88.27%, outperforming the state of the art accuracy reported by Socher et al. [361] (85.40%). Table 6.6 shows the results with ensemble classification and without ensemble classification.

Table 6.7 presents a comparison of the proposed system with well-known state of the art.

6.4.1.2 Results on the Blitzer-derived Dataset

On the Blitzer-derived dataset described in Sect. 6.3.2.2, an accuracy of 88.35% was achieved at the sentence level (see Table 6.8). We tested the performance of the other benchmark sentiment analysis systems on this dataset and the comparative accuracy is reported in Table 6.9. Further, the method by Socher et al. [361] was found to perform very poorly on the Blitzer dataset.

6.4.1.3 Results on the Amazon Dataset

Table 6.10 presents the results of dynamic sentic patterns on the Amazon dataset described in Sect. 6.3.2.3. Table 6.11 shows comparison with the state-of-the-

Table 6.6 Results obtained using different algorithms on movie review dataset

Algorithm	Precision
Sentic patterns	87.15%
CI classifier	76.50%
Ensemble classification	88.27%

Table 6.7 Comparison with the state of the art on movie review dataset

System	Precision
Socher et al. 2012 [359]	80.00%
Socher et al. 2013 [361]	85.40%
Proposed method	88.27%

Table 6.8 Results obtained using different algorithms on Blitzer-derived dataset

Algorithm	Precision
Sentic patterns	86.46%
CI classifier	79.12%
Ensemble classification	88.35%

Table 6.9 Comparison with the state of the art on Blitzer-derived dataset

System	Precision
Socher et al. 2013 [361]	61.93%
Proposed method	88.35%

Table 6.10 Results obtained using different algorithms on the Amazon dataset

Algorithm	Precision
Sentic patterns	80.62%
CI classifier	73.82%
Ensemble classification	83.10%

Table 6.11 Comparison with state-of-the-art on the Amazon dataset

System	Precision
Socher et al. 2013 [361]	68.21%
Proposed method	83.10%

art. Again, the proposed method was found to outperform the state-of-the-art approaches.

6.4.2 Discussion

The proposed approach outperforms state-of-the-art methods both on the movie review and the amazon dataset and shows even better results on the Blitzer-derived dataset, which proves that our system is robust and not biased towards a particular domain. Moreover, it is important to note that, while standard statistical methods require a lot of training, both in terms of resources (training corpora) and time (learning time), dynamic sentic patterns are mostly unsupervised - except for the supervised classification module.

The addition and improvement of the patterns, as noted in Poria et al. (2014) [292], has helped the system improve its results. Tables 6.7, 7.21, and 6.11 show the performance improvement over Poria et al. (2014) [292]. On the other hand [361] has failed to obtain consistently good accuracy over both Blitzer and amazon datasets but obtained good accuracy over the movie review dataset. This is because the classifier proposed in [361] was trained on the movie review dataset only.

The proposed approach has therefore obtained a better accuracy than the baseline system. We combined the three datasets described in Sects. 6.3.2.1, 6.3.2.2 and 6.3.2.3 to evaluate the sentic patterns. From Sect. 6.3.2, we can calculate the number of positive and negative sentences in the dataset, which shows 72,721 positive and 56,903 negative sentences. If the system predicts all sentences as positive, this would give a baseline accuracy of 56.10%. Clearly, the proposed system performed well above than the baseline system. It is worth noting that the accuracy of the system crucially depends on the quality of the output of the dependency parser, which relies on grammatical correctness of the input sentences. All datasets, however, contain ungrammatical sentences which penalize results.

On the other hand, the formation of a balanced dataset for the supervised classifiers actually has a strong impact on developing an accurate classifier. Next, we briefly discuss the performance of the proposed architecture on various dynamic linguistic patterns and types of sentence structures.

Table 6.12 Performance of the proposed system on sentences with conjunctions and comparison with state-of-the-art

System	AND	BUT
Socher et al. 2013 [361]	84.26%	39.79%
Poria et al. 2014 [292]	87.91%	84.17%
Proposed approach	88.24%	85.63%

6.4.2.1 Effect of Conjunctions

Sentiment is often very hard to identify when sentences have conjunctions. We tested the performance of the proposed system on two types of conjunctions: *and* and *but*. High accuracy was achieved for both conjunctions. However, the accuracy on sentences containing *but* was somewhat lower as some sentences of this type do not match sentic patterns. Just over 27% of the sentences in the dataset have *but* as a conjunction, which implies that the rule for *but* has a very significant impact on the accuracy.

Table 6.12 shows the accuracy of the proposed system on sentences containing *but* and *and*, compared with the state-of-the-art. To the best of our knowledge, no state-of-the-art system reports accuracy on sentences containing *and*. Table 6.12 reports the average accuracy of the systems over all datasets.

6.4.2.2 Effect of Discourse Markers

We used a discourse parser developed by Lin et al. [210] to analyze the discourse structure of sentences. Out of the 1211 sentences in the movie review and the Blitzer dataset that contain discourse markers (*though*, *although*, *despite*), we have correctly identified sentiment in 85.67% sentences. The discourse parser sometimes failed to detect the discourse structure of sentences such as *So, although the movie bagged a lot, I give very low rating*. We overcome such problems by removing the occurrence of any word before the discourse marker when a discourse marker occurs at either second or third position in the sentence.

6.4.2.3 Effect of Negation

Through the linguistic rules described in Sect. 6.2.2, we detected negation and studied its impact on sentence polarity. Overall, we achieved 93.84% accuracy on polarity detection from sentences with negation.

Socher et al. [361] state that negation does not always reverse the polarity. According to their theory, the sentence "I do not like the movie" does not bear any negative sentiment but rather is neutral. Another example: "The movie is not terrible"; their theory suggests that this sentence does not say that the movie is good but rather says that it is less bad, so this sentence bears negative sentiment. However, in our annotation we did not follow this theory. We believe that the expression "not bad" implies satisfaction; thus, we annotated such a sentence as positive. Conversely, "not good" implies dissatisfaction and, thus, bears negative sentiment. Based on this argument, we consider the sentence "The movie is not terrible" to be positive.

6.4.2.4 Examples of Differences Between the Proposed System and State-of-the-Art Approaches

Table 6.13 shows examples of various linguistic patterns and the performance of our system across different sentence structures.

Examples in Table 6.14 show that the proposed system produces consistent results on sentences carrying the same meaning although they use different words. In this example, we change the negative sentiment bearing word in the sentence: in the first variant it is **bad**, in the second variant it is **bored**, and in the third variant it

Table 6.13 Performance comparison of the proposed system and state-of-the art approaches on different sentence structures

Sentence	Socher et al. [361]	Proposed system	Correct polarity
Hate iphone with a passion	Positive	Negative	Negative
Drawing has never been such easy in computer	Negative	Positive	Positive
The room is so small to stay	Neutral	Negative	Negative
The tooth hit the pavement and broke	Positive	Negative	Negative
I am one of the least happy people in the world	Neutral	Negative	Negative
I love starbucks but they just lost a customer	Neutral	Negative	Negative
I doubt that he is good	Positive	Negative	Negative
Finally, for the beginner there are not enough conceptual clues on what is actually going on	Positive	Negative	Negative
I love to see that he got injured badly	Neutral	Positive	Positive
I love this movie though others say it's bad	Neutral	Positive	Positive
Nothing can be better than this	Negative	Positive	Positive
The phone is very big to hold	Neutral	Negative	Negative

Table 6.14 Performance of the system on sentences bearing same meaning with different words

Sentence	Socher et al. [361]	Proposed System	Correct Polarity
I feel **bad** when Messi scores fantastic goals	Neutral	Negative	Negative
I feel **bored** when Messi scores fantastic goals	Negative	Negative	Negative
I feel **upset** when Messi scores fantastic goals	Positive	Negative	Negative
I gave her a **gift**	Neutral	Positive	Positive
I gave her **poison**	Neutral	Negative	Negative

Table 6.15 Results obtained using SentiWordNet

Dataset	Using SenticNet	Using SentiWordNet
Movie review	88.27%	88.15%
Blitzer	88.35%	88.31%
Amazon	83.10%	80.62%

is **upset**. In each case, our system detects the sentiment correctly. This analysis also illustrates inconsistency of state-of-the-art approaches, given that the system [361] achieves the highest accuracy compared with other existing state-of-the-art systems.

6.4.2.5 Results Obtained Using SentiWordNet

We carried out an extensive experiment using SentiWordNet instead of SenticNet on all the three datasets. The results showed SenticNet performed slightly better than SentiWordNet. A possible future direction of this work is the invention of a novel approach to combine SenticNet and SentiWordNet in the sentiment analysis framework. The slight difference in the accuracy reported in Table 6.15 confirmed that both the lexicons share similar knowledge but since SenticNet contains concepts, this helps produce increased accuracy. For example, in the sentence "The battery lasts little", the proposed algorithm extracts the concept "last little" which exists in SenticNet but not in SentiWordNet. As a result, when SenticNet is used the framework labels the sentence with a "negative" sentiment but when using SentiWordNet the sentence is labeled with a "neutral" sentiment.

Examples of Cases when Supervised Classifier was used

Below in Table 6.16 we give some examples where supervised classifier was used to guess the polarity. For each of these sentences, no concept was found in SenticNet.

Table 6.16 Some examples where supervised classifier was used to obtain the polarity label

Sentence	Polarity
I had to return the phone after 2 days of use	Negative
The phone runs recent operating system	Positive
The phone has a big and capacitive touchscreen	Positive
My iphone battery lasts only few hours	Negative
I remember that I slept at the movie hall	Negative

6.5 Conclusion

Between the birth of the Internet and 2003, the year of birth of social networks such as MySpace, Delicious, LinkedIn and Facebook, there were just a few dozen exabytes of information on the Web. Today, the same amount of information is created weekly. The advent of the Social Web has provided people with new dynamic content-sharing services that allow them to create and share their own contents, ideas, and opinions, in a time- and cost-efficient way, with virtually millions of other people connected to the World Wide Web. This huge amount of dynamic information, however, is mainly unstructured (as it is specifically produced for human consumption) and hence not directly machine-processable. The automatic real-time analysis of dynamic text in big, heterogeneous, noisy and ambiguous environments such as the web, involves a deep understanding of natural language by machines—a major goal of computational intelligence from which we are still very far off.

Hitherto, online information retrieval, aggregation, and processing have mainly been based on static algorithms relying on the textual representation of webpages. Such algorithms are very good at retrieving texts, splitting them into parts, checking the spelling and counting the number of words. When it comes to dynamically interpreting sentences and extracting meaningful information, however, their capabilities are known to be very limited.

Dynamic concept-level sentiment analysis can help with this, in which, unlike other word-based approaches, it focuses on a semantic analysis of text through the use of web ontologies or semantic networks, which allow the aggregation of conceptual and affective information associated with natural language opinions. Concept-level sentiment analysis, however, is limited by the richness of the knowledge base and by the fact that the bag-of-concepts model, despite being more sophisticated than bag-of-words, misses out important discourse structure information that is key for effectively and dynamically detecting the polarity conveyed by natural language opinions.

In this work, we developed and applied a novel paradigm to concept-level sentiment analysis that merges linguistics, common-sense computing, and computational intelligence for improving the accuracy of dynamic polarity detection. By allowing sentiments to flow from concept to concept based on the dependency relation of the input sentence, we achieve a better understanding of the contextual role of each concept within the sentence and, hence, obtain a dynamic polarity detection engine that outperforms state-of-the-art statistical methods.

There are a number of possible extensions of this work. One is to further develop sentic patterns, which we have shown to play a key role in dynamic concept-level sentiment analysis. In particular, we plan to carry out additional experiments using diverse datasets to further evaluate the domain independence capabilities of linguistic patterns, and compare their performance with other benchmark state-of-the-art approaches (such as [28]). Further, whilst our presumption is that all linguistic patterns are equally important for calculating the polarity of natural language sentences, it would be interesting to carry out further detailed theoretical analysis to investigate relative contributions of our proposed dynamic patterns across a range of datasets from different domains.

Another direction is to expand the common-sense knowledge base, as well as the accuracy of discourse, dependency parsing techniques and the use of probabilistic generative models [200] to improve the accuracy of the framework.

As discussed in Sect. 6.4.1, sarcasm detection plays an important role in sentiment analysis, as the presence of irony usually flips the polarity of a sentence. We plan to develop a sarcasm detection module trained on available datasets[3] [116]. Identifying implicit negation [320] when irony is detected, is also one of the key future tasks of the proposed framework. Finally, we plan to develop other modules, e.g., microtext analysis and anaphora resolution, as part of our vision to develop a novel holistic approach for solving the multi-faceted dynamic sentiment analysis challenge [47].

[3]http://alt.qcri.org/semeval2015/task11

Chapter 7
Combining Textual Clues with Audio-Visual Information for Multimodal Sentiment Analysis

Abstract In this chapter, we present different methods based on classical machine learning and modern Deep Learning based for multimodal sentiment analysis. To accomplish multimodal fusion for sentiment analysis we extract features from textual, audio and visual modality. The proposed framework outperforms the state-of-the-art model in multimodal sentiment analysis research with a margin of 10–13% and 3–5% accuracy on polarity detection and emotion recognition, respectively. This chapter also proposes an extensive study on decision-level fusion.

Keywords Multimodal sentiment analysis · Convolutional neural network · Deep learning · Sentiment · Emotion · MKL · ELM · SVM · Classification

7.1 Introduction

In the last chapter we discuss the approach to text based sentiment analysis. That discussion includes different feature extraction methods from text modality. In this chapter, we present an approach to combine the textual features with audio-visual information. In particular, we have carried out sentiment analysis on videos. Video contains visual, audio and textual modalities. Experiments show, text based features can largely improve the overall multimodal sentiment analysis performance.

7.1.1 Utterance Level Sentiment Analysis

In linguistics utterance is "an uninterrupted chain of spoken or written language". An opinionated video may contain many of such utterances. Each utterance may express different opinions on the product. In this work, we focus on utterance level opinion mining which grasps the sentiment and emotion dynamics throughout a video. On the other hand, video level opinion mining combines the opinion information of all these utterances in order to compute the overall opinion expressed.

S. Poria et al., *Multimodal Sentiment Analysis*, Socio-Affective Computing 8,
https://doi.org/10.1007/978-3-319-95020-4_7

For text based opinion, in the literature both sentence and document level opinion mining are carried out. This is because the training corpora of the document level opinion mining contains self evaluated tagging i.e. reviewers give a rating to the product they are reviewing. In fact, this facilitates modeling the textual opinion mining framework. However, multimodal opinion corporas do not have this information. Hence, every video should be manually labeled in order to perform video level opinion mining. Needless to say, this manual labeling may be highly biased and create noise as it is not possible to validate that labeling with reviewer's rating to the product.

7.1.2 Datasets Characteristics

In this chapter, MOUD, Youtube and ICT-MMMO datasets (described in Sect. 3.2) have been used for sentiment analysis. In order to detect emotions in videos, IEMOCAP (explained in Sect. 3.2) dataset has been used. Though both of these datasets contain emotionally rich videos, there is a fundamental difference between these two datasets.

MOUD, Youtube and ICT-MMMO dataset contain reviews of the products, movies, books. In these videos reviewers discuss several aspects of the opinion topic. On the other hand, IEMOCAP contains dyadic conversational videos where two people are expressing emotions while talking.

MOUD and Youtube datasets are annotated in utterance level. On the other hand, ICT-MMMO dataset is not annotated in utterance level. As this work primarily focuses on utterance level opinion mining, we manually divided ICT-MMMO datasets in the utterance level (Fig. 7.1).

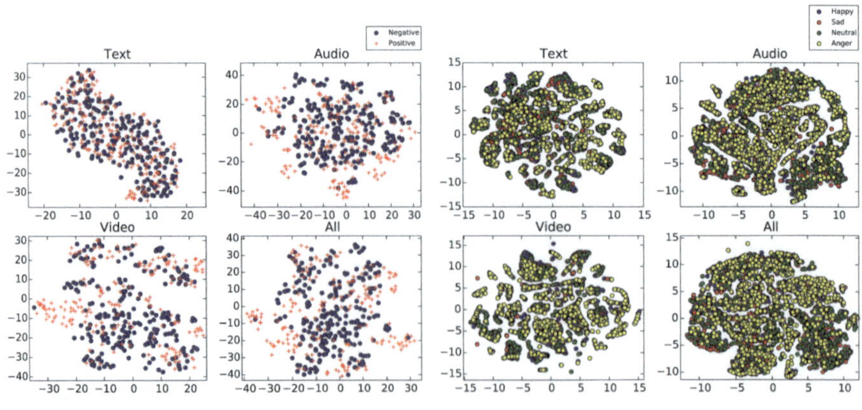

Fig. 7.1 Visualization of MOUD and IEMOCAP datasets when unimodal features and multimodal features are used

7.1.3 Multiple Kernel Learning

Consider a sequence of utterances $s(1), s(2), \ldots, s(T)$. The corresponding features for each utterance from audio, video and text data are denoted by $x(t)^a$, $x(t)^v$ and $x(t)^t$. Multiple kernel learning uses the corresponding target labels $y(t) \in \{+ve, -ve\}$ to optimize a dual form objective function with both min and max terms:

$$\max_{\beta} \min_{\alpha} \frac{1}{2} \sum_{i=1}^{T} \sum_{j=1}^{T} \alpha_i \alpha_j y(i) y(j) \left(\sum_{m=1}^{M} \beta_m^a K_m^a(x(i)^a, x(j)^a) + \sum_{m=1}^{M} \beta_m^v K_m^v(x(i)^v, x(j)^v) + \sum_{m=1}^{M} \beta_m^t K_m^t(x(i)^t, x(j)^t) \right) - \sum_{i=1}^{T} \alpha_i,$$

$$\text{s.t} \sum_{i=1}^{T} \alpha_i y(i) = 0, \ \sum_{m=1}^{\text{M}} \beta_m = 1, 0 \le \alpha_i \le C \forall i. \quad (7.1)$$

where M is the total number of positive definite Gaussian kernels $K_m^a(x(i)^a, x(j)^a)$, $K_m^v(x(i)^v, x(j)^v)$ and $K_m^t(x(i)^t, x(j)^t)$ in each modality with a set of different parameters and α_i, b and $\beta_m \ge 0$ are co-efficients to be learned simultaneously from the training data using quadratic programming.

7.2 Extracting Features from Textual Data

In Chaps. 4, 5, and 6, we have discussed the text sentiment analysis modules. We integrate all of them together and fuse with the visual and audio information in order to multimodal sentiment analysis. For better understanding of the text feature extraction process, we summarize the features and extraction process below –

- **Deep Textual Features –** A deep CNN has been employed on the utterances in order to extract textual features. This CNN structure is discussed in Sect. 6.3.1.1. The total number of features extracted are 500.
- **Common-Sense Knowledge Features –** Common-sense knowledge features consist of concepts are represented by means of EmoSenticSpace. In particular, concepts extracted from text through the semantic parser (as explained in Sect. 4.2) are encoded as 100-dimensional real-valued vectors and then aggregated into a single vector representing the sentence by coordinate-wise summation:

$$x_i = \sum_{j=1}^{N} x_{ij},$$

where x_i is the i-th coordinate of the sentence's feature vector, $i = 1, \ldots, 100$; x_{ij} is the i-th coordinate of its j-th concept's vector, and N is the number of concepts in the sentence. EmoSenticSpace is thoroughly presented and explained in Chap. 5.

- **Sentic Feature –** The polarity scores of each concept extracted from the sentence were obtained from SenticNet and summed up to produce a single scalar feature.
- **Modification Feature –** This is a single binary feature. For each sentence, we obtained its dependency tree from the dependency parser. This tree was analyzed to determine whether there is any word modified by a noun, adjective, or adverb. The modification feature is set to 1 in case of any modification relation in the sentence; 0 otherwise.
- **Negation Feature –** Similarly, the negation feature is a single binary feature determined by the presence of any negation in the sentence. It is important because the negation can invert the polarity of the sentence.

7.3 Extracting Features from Visual Data

Humans are known to express emotions through facial expression, to a great extent. As such, these expressions play a significant role in the identification of emotions in a multimodal stream. A facial expression analyzer automatically identifies emotional clues associated with facial expressions, and classifies these expressions to define sentiment categories and discriminate between them. We use positive and negative as sentiment classes in the classification problem. In the annotations provided with the datasets we use, each video was segmented into utterances and each of the utterances has the length of a few seconds. Every utterance was annotated as either 1, 0 and -1, denoting positive, neutral and negative sentiment.

Using a matlab code, we converted all videos in the dataset to image frames, afterwhich we extracted facial features from each image frame. To extract facial characteristic points (FCPs) from the images, we used the facial recognition library CLM-Z [16]. From each image we extracted 68 FCPs; see examples in Table 7.1. The FCPs were used to construct facial features, which were defined as distances between FCPs; see examples in Table 7.2.

GAVAM[331] was also used to extract facial expression features from the face. Table 7.3 shows the extracted features from facial images. In our experiment we used the features extracted by CLM-Z along with the features extracted using GAVAM.

If a segment of a video has n number of images, then we extracted features from each image and take mean and standard deviation of those feature values in order to compute the final facial expression feature vector for an utterance.

Table 7.1 Some relevant facial characteristic points (out of the 68 facial characteristic points detected by CLM-Z)

Features	Description
48	Left eye
41	Right eye
43	Left eye inner corner
46	Left eye outer corner
47	Left eye lower line
44	Left eye upper line
40	Right eye inner corner
37	Right eye outer corner
42	Right eye lower line
38	Right eye upper line
23	Left eyebrow inner corner
25	Left eyebrow middle
27	Left eyebrow outer corner
22	Right eyebrow inner corner
20	Right eyebrow middle
18	Right eyebrow outer corner
52	Mouth top
58	Mouth bottom
55	Mouth left corner
49	Mouth right corner
14	Middle of the left mouth side
4	Middle of the right mouth side

Table 7.2 Some important facial features used for the experiment

Features
Distance between right eye and left eye
Distance between the inner and outer corner of the left eye
Distance between the upper and lower line of the left eye
Distance between the left iris corner and right iris corner of the left eye
Distance between the inner and outer corner of the right eye
Distance between the upper and lower line of the right eye
Distance between the left eyebrow inner and outer corner
Distance between the right eyebrow inner and outer corner
Distance between top of the mouth and bottom of the mouth
Distance between left and right mouth corner
Distance between the middle point of left and right mouth side
Distance between lower nose point and upper mouth point

7.4 Extracting Features from Audio Data

We automatically extracted audio features from each annotated segment of the videos. Audio features were also extracted in 30 Hz frame-rate and we used a sliding window of 100 ms. To compute the features we used the open source

Table 7.3 Features extracted using GAVAM from the facial features

Features
The time of occurrence of the particular frame in milliseconds
The displacement of the face w.r.t X-axis. It is measured by the displacement of the normal to the frontal view of the face in the X-direction
The displacement of the face w.r.t Y-axis
The displacement of the face w.r.t Z-axis
The angular displacement of the face w.r.t X-axis. It is measured by the angular displacement of the normal to the frontal view of the face with the X-axis
The angular displacement of the face w.r.t Y-axis
The angular displacement of the face w.r.t Z-axis

software openSMILE [112]. Specifically, this toolkit automatically extracts pitch and voice intensity. Z-standardization was used to perform voice normalization. Basically, voice normalization was performed and voice intensity was thresholded to identify samples with and without voice. The features extracted by openSMILE consist of several Low Level Descriptors (LLD) and their statistical functionals. Some of the functionals are *amplitude mean, arithmetic mean, root quadratic mean, standard deviation, flatness, skewness, kurtosis, quartiles, inter-quartile ranges, linear regression slope* etc. Taking into account all functionals of each LLD, we obtained 6373 features. Some of the useful key LLD extracted by openSMILE are described below.

- Mel frequency cepstral coefficients – MFCC were calculated based on short time Fourier transform (STFT). First, log-amplitude of the magnitude spectrum was taken, and the process was followed by grouping and smoothing the fast Fourier transform (FFT) bins according to the perceptually motivated Mel-frequency scaling.
- Spectral Centroid – Spectral Centroid is the center of gravity of the magnitude spectrum of the STFT. Here, $M_i[n]$ denotes the magnitude of the Fourier transform at frequency bin n and frame i. The centroid is used to measure the spectral shape. A higher value of the centroid indicates brighter textures with greater frequency. The spectral centroid is calculated as

$$C_i = \frac{\sum_{i=0}^{n} n M_i[n]}{\sum_{i=0}^{n} M_i[n]}$$

- Spectral Flux – Spectral Flux is defined as the squared difference between the normalized magnitudes of successive windows:

$$F_i = \sum_{n=1}^{n} (N_t[n] - N_{t-1}[n])^2$$

where $N_t[n]$ and $N_{t-1}[n]$ are the normalized magnitudes of the Fourier transform at the current frame t and the previous frame $t-1$, respectively. The spectral flux represents the amount of local spectral change.

- Beat histogram – It is a histogram showing the relative strength of different rhythmic periodicities in a signal. It is calculated as the auto-correlation of the RMS.
- Beat sum – This feature is measured as the sum of all entries in the beat histogram. It is a very good measure of the importance of regular beats in a signal.
- Strongest beat – It is defined as the strongest beat in a signal, in beats per minute, and it is found by identifying the strongest bin in the beat histogram.
- Pause duration – Pause direction is the percentage of time the speaker is silent in the audio segment.
- Pitch – It is computed by the standard deviation of the pitch level for a spoken segment.
- Voice Quality – Harmonics to noise ratio in the audio signal.
- PLP – The Perceptual Linear Predictive Coefficients of the audio segment were calculated using the openSMILE toolkit.

7.5 Experimental Results

In this section, I discuss the experimental results obtained on the MOUD dataset. For the experiment, we removed all neutral classes resulting in the final dataset of 448 utterances. Of these, 247 were negative and 201 were positive. In this section, we describe the experimental results of the unimodal and multimodal frameworks. For each experiment, we carried out 10-fold cross validation.

7.5.1 Extracting Sentiment from Visual Modality

To extract sentiment from only visual modality we used SVM classifier with a polykernel. Features were extracted using the method explained in Sect. 7.3. Table 7.4 shows the results for each class –{positive and negative}.

Clearly, the recall is lower for positive samples. This means many negative instances were labelled as positive. Below, we show some features which took major role to confuse the classifier.

Table 7.4 Confusion matrix for the visual modality (SVM classifier)

Actual	Predicted classification			
classification	Negative	Positive	Precision	Recall
Negative	197	50	76.40%	79.80%
Positive	61	140	73.70%	69.70%

Fig. 7.2 A sample of facial characteristic points extracted by CLM-Z

- The large change in distance of FCPs on eyelid from lower eyebrow.
- Small change between the two corners of the mouth (F_{49} and F_{55} as shown in Fig. 7.2).

We compared the performance of SVM with other classifiers like Multilayer Perceptron (MLP) and Extreme Learning Machine (ELM) [152, 153]. SVM was found to produce best performance results. On visual modality, the best state-of-the-art result on this dataset was obtained by [324] where they got 67.31% accuracy. In terms of accuracy our method has outperformed their result by achieving 75.22% accuracy.

7.5.2 *Extracting Sentiment from Audio Modality*

For each utterance, we extracted the features as stated in Sect. 3.3.2 and formed a feature vector which was then fed to SVM. Table 7.5 shows that for the positive class, the classifier obtained relatively lower recall than for the visual modality obtained. Rosas et al. [324] obtained 64.85% accuracy on audio modality. Conversely, a 74.49% accuracy was obtained using the proposed method, outperforming the accuracy of the state-of-the-art-model [324]. For 1 utterance in the dataset, there is no audio data. This resulted in 447 utterances in the final dataset for this experiment.

Table 7.5 Confusion matrix for the audio modality (SVM classifier)

Actual classification	Predicted classification			
	Negative	Positive	Precision	Recall
Negative	208	38	73.20%	84.60%
Positive	76	125	76.70%	62.20%

Table 7.6 Confusion matrix for the textual modality (CNN classifier)

Actual classification	Predicted classification			
	Negative	Positive	Precision	Recall
Negative	210	36	78.65%	85.36%
Positive	57	143	79.88%	71.50%

7.5.3 *Extracting Sentiment from Textual Modality*

As we described in Sect. 3.2.3, deep Convolutional Network (CNN) was used to extract features from textual modality and a SVM classifier was then employed on those features to identify sentiment. We call this hybrid classifier CNN-SVM. Comparing the performance of CNN-SVM with other supervised classifiers, we found it to offer the best classification results (Table 7.6). In this experiment, our method also outperformed the state-of-the-art accuracy achieved by [324]. For 2 utterances, no text data was available in the dataset. So, the final dataset for this experiment consists of 446 utterances out of which 246 are negative and 200 are positive.

The results shown in Table 7.6 were obtained when the utterances in the dataset were translated from Spanish to English. Without this translation process we obtained a much lower accuracy of 68.56%. Another experimental study showed that while using CNN-SVM produced a 79.14% accuracy, an accuracy of only 75.50% was achieved using CNN.

7.5.4 *Feature-Level Fusion of Audio, Visual and Textual Modalities*

After extracting features from all modalities, we merged them to form a long feature vector. That feature vector was then fed to MKL for the classification task. We tested several polynomial kernels of different degree and RBF kernels having different gamma values as base kernels in MKL. We compared the performance of SPG-GMKL (Spectral Projected Gradient-Generalized Multiple Kernel Learning) [161] and Simple-MKL in the classification task and found that SPG-GMKL outperformed Simple-MKL with a 1.3% relative error reduction rate. Based on the cross validation performance, the best set of kernels and their corresponding parameters were chosen. Finally, we chose a configuration with 8 kernels: 5 RBF with gamma from 0.01 to 0.05 and 3 polynomial with powers 2, 3, 4.

Table 7.7 Confusion matrix for the audio-visual modality (MKL classifier)

Actual classification	Predicted classification			
	Negative	Positive	Precision	Recall
Negative	214	32	82.90%	87.00%
Positive	44	156	83.00%	78.00%

Table 7.8 Confusion matrix for the audio-textual modality (SPG-GMKL classifier)

Actual classification	Predicted classification			
	Negative	Positive	Precision	Recall
Negative	217	29	83.46%	88.21%
Positive	43	157	84.40%	78.50%

Table 7.9 Confusion matrix for the visual-textual modality (MKL classifier)

Actual classification	Predicted classification			
	Negative	Positive	Precision	Recall
Negative	221	25	84.03%	89.83%
Positive	42	158	86.33%	79.00%

Table 7.10 Confusion matrix for the audio-visual-textual modality (SPG-GMKL classifier)

Actual classification	Predicted classification			
	Negative	Positive	Precision	Recall
Negative	227	19	86.64%	92.27%
Positive	35	165	89.67%	82.50%

Table 7.7 shows the results of the audio-visual feature-level fusion. Clearly, the performance in terms of both precision and recall increased when these two modalities are fused.

Among the unimodal classifiers, textual modality was found to provide the most accurate classification result. We observed the same fact when textual features were fused with audio and visual modalities. Both the audio-textual (Table 7.8) and visual-textual (Table 7.9) framework outperformed the audio-visual framework. According to the experimental results, visual-textual modality performed best (Table 7.10).

Table 7.11 shows the results when all three modalities were fused producing a 87.89% accuracy. Clearly, this accuracy is higher than the best state-of-the-art framework, which obtained a 74.09% accuracy. The fundamental reason for our method outperforming the state-of-the-art method is the extraction of salient features from each modality before fusing those features using MKL.

7.5.4.1 Feature Selection

In order to see whether a reduced optimal feature subset can produce a better result than using all features, we conducted a cyclic Correlation-based Feature Subset Selection (CFS) using the training set of each fold. The main idea of CFS is that

Table 7.11 Results and comparison of unimodal experiment and multimodal feature-level fusion (Accuracy)

	Perez-Rosa et al. [324]	Our method
Audio modality	64.85%	74.49%
Visual modality	67.31%	75.22%
Textual modality	70.94%	79.14%
Visual and text-based features	72.39%	84.97%
Visual and audio-based features	68.86%	82.95%
Audio and text-based features	72.88%	83.85%
Fusing all three modalities	74.09%	87.89%

Table 7.12 Results and comparison of unimodal experiment and multimodal feature-level fusion (Accuracy): feature selection was carried out

	Perez-Rosa et al. [324]	Our method
Audio modality	64.85%	74.22%
Visual modality	67.31%	76.38%
Textual modality	70.94%	79.77%
Visual and text-based features	72.39%	85.46%
Visual and audio-based features	68.86%	83.69%
Audio and text-based features	72.88%	84.12%
Fusing all three modalities	74.09%	88.60%

useful feature subsets should contain features that are highly correlated with the target class while being uncorrelated with each other. However, superior results were obtained when we used all features. This signifies that some relevant features were excluded by CFS. We then employed Principal Component Analysis (PCA) for feature selection to rank all features according to their importance in classification. To measure whether top K features selected PCA can produce better accuracy, we fed the top K features to the classifier. However, even worse accuracy was obtained than when using CFS based feature selection. When we took the combination of top K features from that ranking and CFS-based selected features and employed the classifier on them, we observed the best accuracy. To set the value of K, an exhaustive search was made and finally we found that $K = 300$ gave the best result. This evaluation was carried out for each experiment stated in Sects. 7.5.1, 7.5.2 and 7.5.4.

For our audio, visual and textual fusion experiment using CFS and PCA, a total 437 features were selected out of which 305 features were textual, 74 were visual and 58 were from audio modality. This proves the fact that textual features were the most important for trimodal sentiment analysis thanks to CNN feature extractor. Table 7.12 shows the comparative evaluation using feature selection method.

7.5.5 Feature-Level Fusion for Multimodal Emotion Recognition

Besides doing the experiment on multimodal sentiment analysis dataset, we also carried out an extensive experiment on multimodal emotion analysis dataset IEMOCAP (Sect. 3.2). We followed the same method as applied for the sentiment analysis dataset. However, instead of taking it as a binary classification task, we considered it as a 4-way classification. This dataset already provides the facial points detected by the markers and we used those features along with the CLM-Z features. There is a particular for doing this – for many images in this dataset the faces are too small to detect by CLM-Z. Using a similar feature selection algorithm as described in Sect. 7.5.4.1, a total of 693 features were selected, of which 85 features were textual, 239 were audio and 369 were from visual modality.

In Table 7.13 we see that both precision and recall of the Happy class is higher. However, Angry and Sad classes are very tough to distinguish from the textual clues. One of the possible reasons is both of these classes are negative emotions and many words are commonly used to express both of the emotions. On the other hand, the classifier was confused and often classified Neutral with Happy and Anger. Interestingly, it classifies Sad and Neutral classes well.

In the case of Audio modality (Table 7.14) we observe better accuracy than textual modality for Sad and Neutral classes. However, for Happy and Angry, the performance decreased. The confusion matrix shows the classifier performed poorly when distinguishing Angry from Happy. Clearly, audio features are unable to effectively classify these based on extracted features. However, the classifier performs very well to discriminate between the classes of Sad and Anger. Overall

Table 7.13 Confusion matrix for the textual modality (SVM classifier, feature selection carried out)

Actual classification	Predicted classification					
	Angry	Happy	Sad	Neutral	Precision	Recall
Angry	650	82	165	186	55.13%	60.01%
Happy	193	957	149	331	68.40%	58.71%
Sad	139	87	619	238	55.51%	57.15%
Neutral	197	273	182	1031	57.72%	61.25%

Table 7.14 Confusion matrix for the audio modality (SVM classifier, feature selection carried out)

Actual classification	Predicted classification					
	Angry	Happy	Sad	Neutral	Precision	Recall
Angry	648	137	89	209	61.53%	59.83%
Happy	159	926	123	422	65.21%	56.81%
Sad	84	152	658	189	63.08%	60.75%
Neutral	162	205	173	1143	58.22%	67.91%

Table 7.15 Confusion matrix for the visual modality (SVM classifier, feature selection carried out)

Actual classification	Predicted classification					
	Angry	Happy	Sad	Neutral	Precision	Recall
Angry	710	83	116	174	66.17%	65.55%
Happy	102	1034	148	346	72.76%	63.43%
Sad	123	83	726	151	63.18%	67.03%
Neutral	138	221	159	1165	63.45%	69.22%

Table 7.16 Confusion matrix for the audio-visual-textual modality (SPG-GMKL classifier, feature selection carried out)

Actual classification	Predicted classification					
	Angry	Happy	Sad	Neutral	Precision	Recall
Angry	821	79	93	90	69.16%	75.80%
Happy	119	1217	92	202	80.11%	74.67%
Sad	93	82	782	126	67.24%	72.20%
Neutral	154	141	196	1192	73.99%	70.82%

identification accuracy of the Neutral emotion has also increased. But Happy and Neutral emotions are still very hard to classify effectively by Audio classifier alone.

Visual modality produced the best accuracy (Table 7.15) when compared to other two modalities. The similar trend has been observed as textual modality. Angry and Sad faces are hard to classify using visual clues. However, Angry and Happy, Happy and Sad faces can be effectively classified. Neutral classes were also separated accurately in respect to other classes.

When we fuse the modalities using the feature-level fusion strategy (Table 7.16) as stated in Sect. 7.5.4, as expected higher accuracy was obtained than with unimodal classifiers. Although the identification accuracy has been improved for every emotion, the confusion between a Sad and Angry face is still higher. Neutral and Sad emotions are also more difficult to classify.

The comparison with the state-of-the-art model in terms of weighted accuracy shows that the proposed method performs significantly better. Comparing the weighted accuracy (WA) with the state of the art, the proposed method obtained 3.75% higher accuracy. However, for Anger emotion class, an approximately 3% lower accuracy was achieved.

7.5.6 Decision-Level Fusion

In this section, we describe different frameworks that we developed for the decision-level fusion. Clearly, the motivation for developing these frameworks is to perform the fusion process in less time. The fusion frameworks were developed according

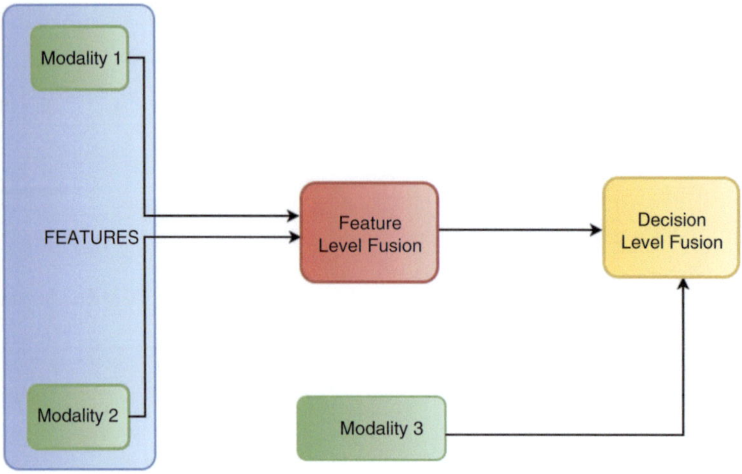

Fig. 7.3 Decision-level fusion framework

to the architecture as shown in Fig. 7.3. Each of the experiments stated below were processed through the feature selection algorithm stated in Sect. 7.5.4.1.

Each block M_i denotes a modality. As the architecture shows, modality M_1 and M_2 are fused using feature-level fusion and then at last stage are fused with another modality M_3 using decision-level fusion. For feature-level fusion of M_1 and M_2, we used SPG-GMKL. The decision-level algorithm is described below.

In decision-level fusion, we obtained the feature vectors from the above-mentioned methods but used separate classifier for each modality instead of concatenating feature vectors as in feature-level fusion. The output of each classifier was treated as a classification score. In particular, from each classifier we obtained a probability score for each sentiment class. In our case, as there are two sentiment classes, we obtained 2 probability scores from each modality. Let, q_1^{12} and q_2^{12} are the class probabilities resulted from the feature level fusion of M_1 and M_2. On the other hand let, q_1^3 and q_2^3 are the class probabilities of modality M_3. We then form a feature vector by concatenating these class probabilities.

We also used sentic patterns as described in Chap. 6 to obtain the sentiment label for each text. If the result by sentic patterns for a sentence is "positive" then we included 1 in the feature vector, otherwise 0 was included in the feature vector. So, the final feature vector looks like this – $[q_1^{12}, q_2^{12}, q_1^3, q_2^3$, sentic] where sentic = 1 if the output of sentic patterns is positive otherwise we set sentic = 0. We then employed SVM on this feature vector in order to obtain the final polarity label.

The best accuracy was obtained when we early fused visual and audio modalities. However, when we fuse all the modalities without carrying out the early fusion, the obtained accuracy was lower. Table 7.17 shows the decision-level accuracy in detail.

Table 7.17 Decision-level fusion accuracy

M_1	M_2	M_3	Sentic patterns	Accuracy
Visual	Audio	Textual	No	73.31%
Visual	Audio	Textual	Yes	78.30%
Visual	Textual	Audio	No	76.62%
Audio	Textual	Visual	No	72.50%

Table 7.18 Decision-level fusion accuracy for multimodal emotion analysis

M_1	M_2	M_3	Weighted accuracy
Visual	Audio	Textual	64.20%
Visual	Textual	Audio	62.75%
Audio	Textual	Visual	61.22%

Table 7.19 Comparison with the state of the art [327] on IEMOCAP dataset

	Rozic et al. [327]	Proposed method
Anger	78.10%	75.80%
Happy	69.20%	74.67%
Sad	67.10%	72.20%
Neutral	63.00%	70.82%
WA	69.50%	73.25%

7.5.6.1 Decision-Level Fusion for Multimodal Emotion Detection

Like decision-level fusion for multimodal sentiment analysis, similar method was applied for multimodal emotion analysis as well (Table 7.18).

Similarly as we saw in the sentiment analysis experiment, the configuration yielding best accuracy was obtained using M_1, M_2 and M_3 as Visual, Audio and Textual respectively.

Table 7.19 shows the detail result of decision-level fusion experiment on IEMOCAP dataset. It should be noted that Sentic Patterns algorithm can not be used in this experiment as it is specific to sentiment analysis.

7.6 Speeding Up the Computational Time: The Role of ELM

In this section the performance obtained using ELM and its comparison with SVM is discussed.

7.6.1 Experiment and Comparison with SVM

The experimental results in Table 7.20 shows ELM and SVM offering equivalent performance in terms of accuracy. While for multimodal sentiment analysis SVM outperformed ELM with a sharp 1.23% accuracy margin, on the emotion analysis

Table 7.20 Accuracy comparison between SVM and ELM (A = Audio, V = Video, T = Textual, UWA = Un-weighted average)

Dataset	A		V		T		A+V+T (UWA)	
	SVM	ELM	SVM	ELM	SVM	ELM	SPG-GMKL	MK-ELM
YouTube	74.22%	73.81%	76.38%	76.24%	79.77%	78.36%	88.60%	87.33%
IEMOCAP	61.32%	60.85%	66.30%	64.74%	59.28%	59.87%	73.37%	72.68%

Table 7.21 Computational time comparison between SVM and ELM

	YouTube dataset	IEMOCAP dataset
SPG-GMKL	1926 s	4389 s
MK-ELM	584 s	2791 s

dataset their performance difference is not significant. On the IEMOCAP dataset, ELM showed better accuracy for text based emotion detection. Importantly, for the purposes of feature-level fusion, we used a multiple kernel variant of the ELM algorithm namely **Multiple Kernel Extreme Learning Machine**. The details of the Multiple Kernel ELM can be found here [216]. As MK-ELM [216] is not in the scope of this work, we encourage readers to refer to that paper. As with SPG-GMKL for feature-level fusion (Sect. 7.5.4), the same set of kernels was used for MK-ELM.

However, ELM edges SVM out by a big margin when it comes to computational time, i.e., training time of feature-level fusion (see Table 7.21).

SPG-GMKL outperformed SVM for the future level fusion task by 2.7%.

7.7 Improved Multimodal Sentiment Analysis: Deep Learning-Based Visual Feature Extraction

In this section, a deep neural network based model is proposed for extracting features from visual modality. This new types of features are found to be more effective than the visual features as explained in Sect. 7.3. For text and audio feature extraction, the methodology as explained in Sects. 6.3.1.1 and 7.4 were used. We begin by discussing Recurrent Neural Networks.

7.7.1 Recurrent Neural Networks

The standard recurrent neural network (RNN) output, $\mathbf{x}_l(t)$, at time step t for each layer l is calculated using the following equations:

$$\mathbf{x}_l(t) = f(W_{\mathrm{R}}^{l}.\mathbf{x}_l(t-1) + W_l.\mathbf{x}_{l-1}(t)) + W_{\mathrm{C}}\int_{t-k}^{t} \mathrm{x}_l(t)dt \tag{7.2}$$

where W_R is the interconnection matrix among hidden neurons and W_l is the weight matrix of connections between hidden neurons and the input nodes, $\mathbf{x}_{l-1}(t)$ is the input vector at time step t from layer $l-1$, vectors $\mathbf{x}_l(t)$ and $\mathbf{x}_l(t-1)$ represent hidden neuron activation at time steps t and $t-1$, respectively, and f is the non-linear activation function. Further, the distributed delays between output hidden features in each layer can be modelled via W_C. Unlike discrete time delays that can be learned separately for each hidden neuron, the distributed time delays are continuously changing due to the combined effect of different outputs and hence we use integration with respect to time to compute them.

In this work, we propose to learn distributed time-delayed dependence using convolution neural networks. Hence, a kernel of dimension $k \times k$ is able to capture distributed delays of upto k time points in the video sequence and can be approximated by the covariance matrix of features learned in the penultimate layer using (7.2). To learn the weights W_R of the RNN, back propagation through time is used where the hidden layer is unfolded in time using duplicate hidden neurons.

7.8 Convolutional Recurrent Multiple Kernel Learning (CRMKL)

In this section, we propose a temporal convolutional neural network (CNN) where each pair of images at time t and $t+1$ are combined into a single image. Such a model is sensitive to sequence of images and learns a dictionary of features that are portable across languages. In a deep CNN, each hidden layer is obtained by convolving a matrix of weights with the matrix of activations at the layer below and the weights are trained using back propagation [198].

Furthermore, we have additional layers of recurrent neurons in the deep model. Recurrent neural networks (RNN) have feedback connections among neurons that can model dependencies in time sequences. Here, each hidden layer state is a function of the previous state, which can be further expanded as a function of all the previous states. In [428], the authors proposed convolutional RNNs to capture spatial structure information in static images. In contrast, our model uses RNN to capture spatial and temporal patterns that are inherent in video sequence. Our experiments showed that while using only RNN or deep CNN does not provide good classifications, combining the two models results in tremendous speed up and accuracy.

Multiple kernel learning (MKL) is a feature selection method where features are organized into groups and each group has its own kernel function [367]. MKL further improved our results, as it is able to combine data from different modalities effectively. Figure 7.4 illustrates the convolutional recurrent multiple kernel learning (CRMKL) model, which combines sentiment features in audio, video and text. In [181], the authors propose the use of a multi-resolution CNN to capture temporal features in YouTube videos. However, to our knowledge, this type of temporal CNN

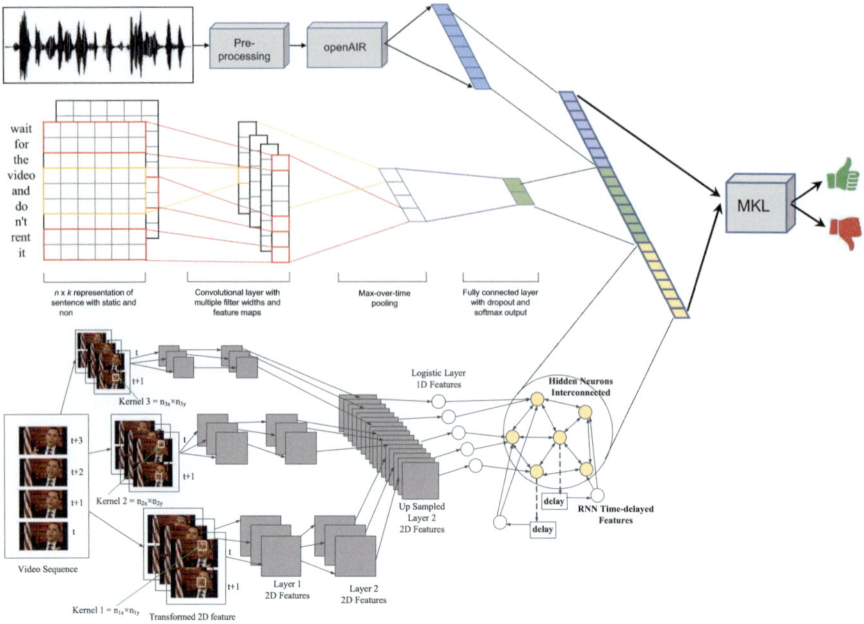

Fig. 7.4 The CRMKL model combining sentiment features in audio, video and text

has not been previously used for sentiment analysis. The following is a summary of the significance and contributions of this research presented in this section:

- We combine video, audio and text modality in order to effectively detect sentiment in a subject-independent manner. Our first contribution is that we have used MKL to fuse the three modalities. While the state of the art [273] uses a single kernel support vector machine (SVM) classifier to fuse all three modalities, we use multiple kernels to adapt to different modalities and, hence, achieve higher accuracy.
- Our second contribution is the novel integration of CNN with a low dimensional RNN, which is computationally much faster on large video data compared to baselines. In particular, for better modeling overlaps among features learned during temporal convolution, we consider distributed time-delayed features in the video. This can be achieved by initializing the weights of RNN with the covariance matrix of output feature vectors learned by the CNN.

7.8.1 Extracting Features from Visual Data

Sentiment analysis of large scale visual content can help to correctly extract sentiment of a topic. Deep convolutional neural networks have good accuracy on

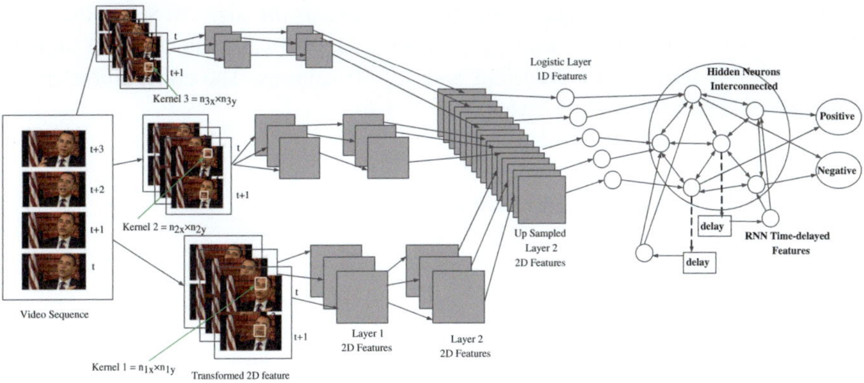

Fig. 7.5 Convolutional recurrent neural network (CRNN) for visual sentiment detection

topic classification of videos, however they can get stuck in a local minima on fine-grained problems such as sentiment and emotion detection. They are also extremely slow. Hence, we propose a layer of recurrent neurons to optimize the learning of features from video data.

Video sentiment detection faces two main challenges, first it is an extremely computationally expensive task and secondly the dataset is weakly labelled and hence the trained model may not generalize well on new datasets. Since, the video data is very large, we only consider every 10th frame in our training videos. The Constrained local model (CLM) is used to find the outline of the face in each frame [17]. The cropped frame size is further reduced by scaling down to a lower resolution. Hence, in this way we can drastically reduce the amount of training video data.

Figure 7.5 illustrates a convolutional recurrent neural network for visual sentiment detection. The input is a sequence of images in a video. To capture the temporal dependence, we transform each pair of consecutive images at t and $t + 1$ into a single image. We use kernels of varying dimensions illustrated as Kernel 1, 2 and 3 to learn Layer 1 2D features from the transformed input. Similarly, the second layer also uses kernels of varying dimensions to learn 2D features. Up sampling layer will transform features of different kernel sizes into uniform 2D features. Next, a logistic layer of neurons is used to prepare input for a RNN. Here, we have an interconnected layer of neurons that can model long time delays using delay states. The final output layer classifies each video image as 'Positive' or 'Negative'.

In order to generalize the model to other domains, we train it using faces of different shapes and sizes. In order to validate it in a speaker independent manner, moreover, we train the model on videos of product reviews in one domain and test on videos from a completely different domain.

Pre-processing involved scaling all video frames to half the resolution. Each pair of consecutive video frames were converted into a single frame so as to achieve temporal convolution features. All the frames were standardized to 250×500 pixels by padding with zeros.

The first convolution layer contains 100 kernels of size 10×20, the next convolution layer had 100 kernels of size 20×30, this was followed by a logistic layer of 300 neurons and a recurrent layer of 50 neurons. The convolution layers were interleaved with pooling layers of dimension 2×2.

7.8.1.1 Feature Selection and Fusion

We have carried out the same feature selection approach as presented in Sect. 7.5.4.1. Feature-level fusion is achieved by concatenation of the feature vectors obtained for each of the three modalities. Clearly, the combined feature vectors from different modalities are heterogeneous in nature. Hence, the resulting vectors, along with the corresponding sentiment polarity labels from the training set, were used to train a classifier with a multiple kernel learning (MKL) algorithm; we used the SPF-GMKL implementation [161], which is designed to deal with heterogeneous data. The parameters of the classifier were found by cross-validation. We chose a configuration with 8 kernels: 5 RBF with gamma from 0.01 to 0.05 and 3 polynomial with powers 2, 3, 4. We also tried Simple-MKL; it gave slightly lower results.

7.8.1.2 Computational Complexity

The computational complexity for a convolutional layer l is given by $O(n_{l-1}.s_l^2.n_l.m_l^2)$, where n_{l-1} and n_l are the number of input and output feature maps, $s_l = n_x^{l-1} \times n_y^{l-1}$ and $m_l = n_x^l \times n_y^l$ are the dimensions of the input and output feature maps. The computational complexity of a layer of recurrent hidden neurons is only $O(\mathrm{R} \times n^2)$, where R is the maximum time delay considered and n is the number of neurons. We can hence conclude that the computational complexity of RNN is much lower than CNN for each iteration of training. Hence, in this work we first train CNN for a limited number of epochs and then the partially learned features are further evolved using a low dimensional RNN for video data.

7.9 Experimental Results and Discussion

We used a common framework for both sentiment and emotion detection. For multimodal sentiment analysis, following Perez et al. we have used the entire set of 448 utterances in MOUD dataset and run ten-fold cross-validation using CRMKL. In addition, to test the generalization ability of the model on new datasets, we have also shown results on test data from YouTube and ICT-MMMO. For comparison with unimodal datasets such as only video or only text, we have used SVM as a baseline classifier. For the case of emotion recognition, that is a much more fine-grained problem than sentiment detection, we evaluate our model via ten-

Table 7.22 Accuracy of state-of-the-art method compared with our method with feature-level fusion on MOUD dataset. The number of features refers to our experiments, not to [273]

		Text	Visual	Audio	[273]	Our method: Without feature selection	Our method: With feature selection
# features, without selection		500	50	6373			
Unimodal	# features,with selection	437	–	–	70.94%	79.14%	79.77%
		–	50	–	67.31%	94.50%	94.50%
		–	–	325	64.85%	74.49%	74.22%
Bimodal		381	50	–	72.39%	95.75%	96.21%
		384	–	81	72.88%	83.85%	84.12%
		–	50	217	68.86%	95.38%	95.68%
Multimodal		50	89	64	74.09%	96.12%	96.55%

fold cross-validation on IEMOCAP dataset. Feature selection was not done for visual modality as the deep CNN module in CRMKL automatically learns the best features. Our experiments showed that feature selection on visual data can lead to reduction in accuracy.

Table 7.22 shows the 10-fold cross-validation results obtained on MOUD dataset. The visual module of CRMKL, obtained 27% higher accuracy than the state of the art. When all modalities were used, 96.55% accuracy was obtained outperforming the state of the art by more than 20%. Next, to assess the accuracy of the model on an unknown dataset, we trained the model on MOUD dataset and tested on ICT-MMMO and YouTube dataset. On both of these datasets, the model performed notably well.

The visual classifier trained on the MOUD obtained 93.60% accuracy. Other unimodal classifiers did not perform well like the visual classifier in the cross-domain analysis. As ICT-MMMO dataset is a video-level sentiment dataset, utterance-level sentiment evaluation is not possible. Hence, after the model generates sentiment labels of all utterances for a video, we took the majority sentiment label of these utterances in order to label the video by its sentiment.

We got 85.30% accuracy on the ICT-MMMO dataset using the trained visual sentiment model on the MOUD dataset. The obtained accuracy on the ICT-MMMO dataset was lower than the other two datasets. This is because ICT-MMMO dataset was manually segmented into utterances and, hence, it is likely to have more noise compared to other datasets.

Not only the visual features, textual features are also novel as they indeed boosted the accuracy of the experiments where textual modality was involved. The unimodal experiment with only textual features outperformed the performance of the state of the art as shown in Table 7.22. On all three datasets, the visual and textual modalities when combine together produced better accuracy than other bimodal experiments.

For multimodal emotion analysis, we used the same framework as we employed for multimodal sentiment analysis. The accuracy for all unimodal, bimodal and trimodal experiments are significantly better than the state of the art. However, the performance is not as good as the multimodal sentiment analysis experiments. One of the possible reasons for this is the use of same CNN configurations for both visual and textual sentiment feature extraction.

This raises the question of using larger number of neurons and layers in CRMKL for visual and textual emotion feature extraction. This is of course a fundamental task of our future work. The following observations were made from the multimodal emotion analysis experiments:

- We realized that the textual classifier recognized *angry*, *happy* and neutral instances well. However, *angry* and *sad* instances are very tough to distinguish from each other using textual clues. One of the possible reasons is that both classes are negative and many similar words are used to express them.
- In the case of audio modality, we observed better accuracy than textual modality for *sad* and neutral classes but not for *happy* and *angry* classes. The classifier misclassified many *happy* instances into *angry*. However, the classifier performed very well to discriminate between sadness and anger. We also observed that some *happy* instances were classified as neutral.
- Visual modality produced the best accuracy compared to the other two modalities. Though *angry* and *sad* faces can be effectively classified, the classifier showed some confusion between *angry* and *sad* faces. Neutral classes were also separated more accurately in respect to other classes though high confusion was observed between *happy* and *neutral* faces.

When we fused the modalities using feature-level fusion strategy, higher accuracy was obtained as compared to unimodal classifiers, as expected. Although the identification accuracy has been improved for every emotion, the confusion between *sad* and *angry* face was still high.

The comparison with the state of the art (Table 7.23) in terms of accuracy shows that the proposed method performed significantly better. For *sad* and neutral emotion the proposed method outperformed the state of the art by a margin of 8% and 17%, respectively. However, for *angry* and *happy* the performance is just slightly better.

Paired t-test showed statistical significance of all experiments with confidence level 95%. It can be found from the Table 7.23 that visual and textual modalities performed notably better than the state of the art. With the help of these two modalities, the proposed method outperformed the state of the art.

In this work, we proposed the novel integration of CNN with a low dimensional RNN that can converge to the global maxima much faster than baselines. Hence, in Table 7.23, for the emotion *sad*, the performance of visual modality and the bimodal combinations of visual modality with text and audio, respectively, is over 70%. This is due to the superior performance of the proposed video classifier. In contrast, the bimodal combination of audio and text has a 10% lower accuracy that is similar to

Table 7.23 Accuracy on textual (T), visual (V), audio (A) modality and comparison with the state of the art

Modalities		Emotion, on IEMOCAP			
		Angry	*Happy*	*Sad*	Neutral
T	**Our results**	60.01%	58.71%	57.15%	61.25%
	State of the art	63.10%[a]	49.60%[a]	42.00%[a]	39.50%[a]
V	**Our results**	69.50%	67.34%	67.41%	71.55%
	State of the art	41.80%[a]	63.60%[a]	52.60%[a]	47.00%[a]
A	**Our results**	59.83%	56.81%	60.75%	67.91%
	State of the art	66.10%[a]	53.90%[a]	65.50%[a]	58.10%[a]
T + V	**Our results**	74.81%	69.22%	74.85%	77.49%
	State of the art	–	–	–	–
T + A	**Our results**	62.50%	65.21%	63.30%	69.25%
	State of the art	77.80%[a]	63.20%[a]	68.30%[a]	60.40%[a]
V + A	**Our results**	71.86%	69.35%	74.23%	77.58%
	State of the art	–	–	–	–
A + V + T	**Our results**	**79.20%**	**72.22%**	**75.63%**	**80.35%**
	State of the art	78.10%[a]	69.20%[a]	67.10%[a]	63.00%[a]

[a]By [327]

the baseline. The combined multimodal classifier of audio, video and text is slightly better than visual modality. This is because the video classifier dominates over the other two modalities.

Deep CNN have recently shown good performance on audio, video and text classification. Instead of using a single large hidden layer of neurons, deep models have several small layers of hidden neurons. Since each layer is independent, this results in tremendous reduction in complexity. Therefore, in this research we construct a deep CNN for each modality, namely: audio, video, and text.

The groups of features learned by each of the three deep CNN are combined using MKL. In this way, we can reduce the number of input dimensions and group the features for MKL.

From the experimental results it can be seen that deep visual features have further boosted the accuracy of both sentiment analysis and emotion recognition. Due to this performance improvement of 10–15% over CLM-Z based facial features, it can be concluded that features extracted using the deep stacked CNN-RNN network is superior to the features obtained using CLM-Z.

7.9.1 Effect of Number of Hidden Layers

Deep learning is able to approximate very long time-delays in video data via a hierarchy of hidden layers, where the features learned in one layer become the input data to the next layer. To determine the number of hidden layers of recurrent neurons, we consider the root mean square error (MSE) on training data. MSE is the cost function that the deep model is trying to minimize while learning the weights. Hence, this is a suitable metric for the improvement made by each hidden layer in a deep model.

Figure 7.6 reports the decrease in MSE with increasing number of hidden layers for the YouTube test dataset. It was also observed that the variance over 10-fold cross-validation reduces with increasing number of hidden layers. Hence, we can conclude that deep learning is suitable for extracting sentiments and emotions from video data. Since each layer is learned independent of the previous layer, the number of parameters is small and overfitting is avoided.

7.9.1.1 Tuning of Hyper-parameters

As a performance measure, we adopted the F-score. Each dataset is split into training set, validation set, and test set. For all three modalities and for each hidden layer we consider different number of hidden units (i.e., n = 50/200/500/700) and 5000 epochs of CNN training using the Theano based stochastic gradient descent. The number of hidden neurons in each layer is gradually increased until performance saturates due to overfitting. In particular, early overfitting occurs for the MOUD dataset.

Our best results are obtained with an ensemble of CNNs by 10-fold cross-validation that differ in their random initialization and mini-batches of 100 samples. Results on CNN's of various depths and sizes shows that deep CNN outperforms

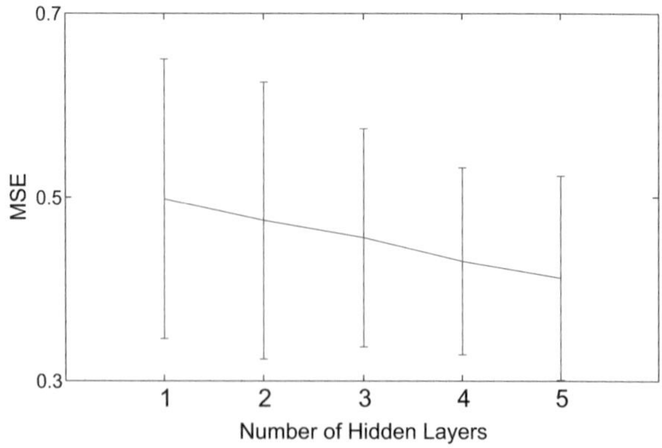

Fig. 7.6 MSE with respect to number of hidden layers

single-layer CNN with approximately the same number of parameters, which quantitatively validates the benefits of deep networks over shallow ones.

We see a consistent improvement as we use deeper models. Following previous authors, the word vector length was empirically set to 300, and unknown words were randomly initialized to vectors from Gaussian distributions. The six dimensional vector corresponds to six different parts of speech such as noun and verb.

7.9.1.2 Visualization of Features

The deep temporal CNN model automatically learns features from the training data, so that each neuron learns a specific feature such as eyes or mouth. In the first layer, the features learned are parts of the face and their sentiments, and the higher layers will combine these emotional features to learn the complete face and corresponding positive or negative sentiment. We visualize the feature detectors in the first layer of the network trained on the MOUD sentiment data. We rank all image segments in the training data according to the activation of each detector. Figure 7.7 shows the top image segments activated at two feature detectors in the first layer of deep CNN. We find that similar features such as eyes or mouth are expressed at the same hidden neuron. The feature detectors learn to recognize not only the part of the face but also the sentiment associated with it.

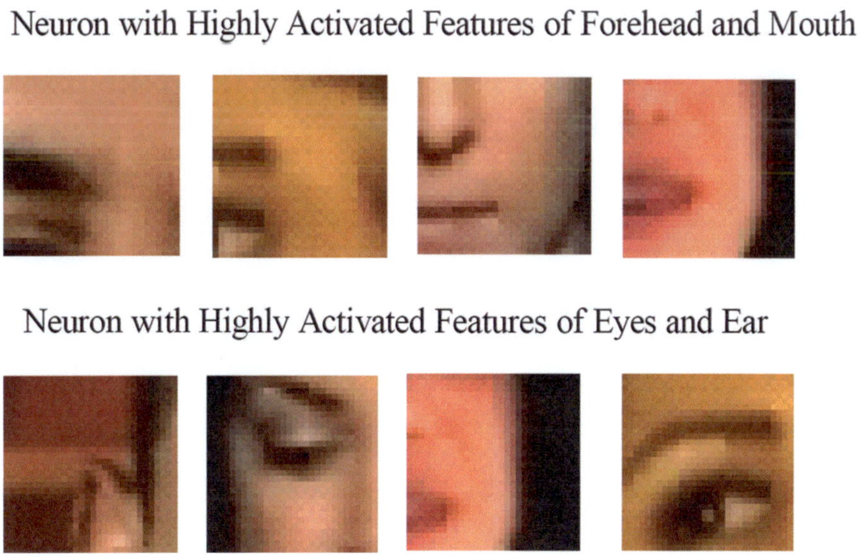

Fig. 7.7 Top image segments activated at two feature detectors in the first layer of deep CNN

7.10 Conclusion

In this chapter, we proposed the fusion of language, voice, and facial expressions for multimodal emotion recognition and sentiment analysis. In particular, we described a novel temporal deep convolutional neural network for visual feature extraction and used multiple kernel learning to fuse heterogeneous features extracted from different modalities, namely: audio, video, and text. The text engines as proposed in the earlier chapters have been integrated in the multimodal system in order improve the system performance. The impact of text modality is found to be consistent on different datasets and tasks. It can be seen from the experimental results that inclusion of text modality has improved the overall recognition rate. On the other hand, one can not ignore the usefulness of the visual modality which has primarily motivated us in developing a novel deep learning based network to extract visual sentiment and emotion features. In the end, we have found that visual features performed better than all other modalities thanks to the effective deep network.

In the future, we will focus on improving the accuracy of emotion detection via different neural network configurations. We will also consider annotation of ICT-MMMO dataset at utterance level for smoother training of the model. We also aim to build a large multimodal sentiment analysis dataset.

Chapter 8
Conclusion and Future Work

Abstract The main aim of this book was to go beyond textual sentiment analysis approaches by integrating audio and visual features with textual for multimodal sentiment analysis. To this end, textual sentiment analysis has also been improved by further developing and applying common-sense computing and linguistic patterns to bridge the cognitive and affective gap between word-level natural language data and the concept-level opinions conveyed by these. Various novel linguistic and machine learning based frameworks have been developed in order to accomplish multimodal sentiment analysis. Apart from the sentiment analysis task, the proposed multimodal model is also capable to detect emotions in videos.

Keywords Multimodal · Multimodal sentiment analysis · Multimodal emotion recognition · Deep learning · LSTM · CNN

Our research is focused on sentence-level sentiment analysis, which we also use to improve utterance-level multimodal sentiment analysis. To this end, we introduced a novel framework, based on both novel and existing machine-learning and knowledge-based techniques.

Concept-based sentiment analysis has generated increasing interest recently. Before any concept-level methods can be applied on text, it is necessary to identify the concepts in text. Therefore, we have designed a concept extraction engine based on the syntactic structure of sentences. Particularly, concepts are extracted based on linguistic patterns which have been designed from dependency trees of sentences.

After extracting concepts from natural language text, for common-sense reasoning and creation of feature space for concept-level sentiment analysis, two resources, namely EmoSenticNet and EmoSenticSpace, have been constructed. A modified fuzzy-SVM method has been employed to construct the EmosenticNet resource. On the other hand, EmoSenticSpace has been developed by first merging EmoSenticNet with Conceptnet and then applying Singular Value Decomposition. As a method of evaluation, we employed 10-fold cross validation, which overcomes the problem of overfitting. EmoSenticNet is a resource which contains concepts and their corresponding emotion, according to Ekman's six basic emotion taxonomy,

S. Poria et al., *Multimodal Sentiment Analysis*, Socio-Affective Computing 8,
https://doi.org/10.1007/978-3-319-95020-4_8

specifically Anger, Disgust, Fear, Sadness, Surprise, and Happiness. EmoSenticSpace is a vector space, where each concept is represented as a d-dimensional vector (typically $d = 100$).

To understand the sentiment expressed in natural language text, we have constructed several linguistic patterns termed as Sentic Patterns, based on the dependency trees of sentences. Experimental results demonstrated that this unsupervised process of employing linguistic patterns in sentiment analysis is very effective. We have fused linguistic patterns with supervised machine learning methods in order to improve the accuracy of sentiment classification. On the benchmark datasets, this joint use of supervised machine learning and linguistic patterns have outperformed the state-of-the-art. Features extracted from EmoSenticSpace, along with other syntactic features have been used in the machine learning classifier. Convolutional Neural Network (CNN) has been used as the supervised machine learning classifier, which has shown to perform significantly better in comparison with other classifiers, such as ELM and SVM.

With the advancement of technology and the rapid rise of social media, along with the large amount of opinions expressed in textual format, there is a growing number of opinions posted in video format. Consumers tend to record their opinions on products via a webcam or other devices and upload them on social media platforms like YouTube or Facebook. This is to inform and educate other people about the products before they make purchase decisions. These videos often contain comparisons of the test product with products from competing brands, the pros and cons of the product, etc. All of this information is useful for people who wish to purchase the product. The main advantage of analyzing videos over solely textual analysis, in order to detect emotions and sentiments from opinions, is the higher availability of cues. However, the impact of textual data in multimodal sentiment analysis has barely been researched in the literature. Only a few studies are available which have considered text modality in multimodal sentiment analysis. Moreover, those works only applied basic Bag Of Words (BOW) based text analysis methods. We overcome this limitation by employing concept-level text sentiment analysis methods like Sentic Patterns, EmoSenticSpace features, and Convolutional Neural Network. In particular, EmoSenticSpace and Convolutional Neural Network help to produce enriched textual features to be used for feature level fusion in the multimodal system. Sentic Patterns have helped in decision level fusion. Inclusion of these novel text-based sentiment analysis methods have improved the overall performance of multimodal sentiment analysis. To this end, we also have developed a novel deep network based on Convolutional Neural Network for visual features extraction, which has further improved the performance of our method. As a method for fusion, we used Multiple Kernel Learning (MKL), which is an extension of the SVM, with more than one kernel. MKL is specifically useful for heterogeneous data, which meets the problem we have dealt with in this research.

8.1 Social Impact

The primary goal of this research is to develop novel techniques for text-based and multimodal sentiment analysis. The research is likely to profoundly impact both industry and consumers. Below, we briefly outline such impacts.

8.1.1 Impact on the Industry

Every company in the industry wants to be aware of the impression of end users on their products. Most companies have a dedicated team to address consumers' feedback. This research will automate feedback and grievance addressing processes. It will determine if consumers like a product, and also extract any suggestion to improve a certain feature of the product (e.g. camera quality of Samsung S6 is bad). This automatic feedback addressal system will greatly benefit companies by identifying buggy or flop features in their products which will further lead to growth of their business.

8.1.2 Impact on the Consumers

On the other hand, consumers will also benefit from this research. Whenever we want to buy a product, we tend to search the internet for consumer reviews. However, manually reading reviews is an unsurmountable task. The proposed text-based sentiment analysis system will automatically mine user reviews of a product and help users make up their mind about the product. This automatic analysis of the reviews has a two fold impact. Firstly, in e-commerce websites like Amazon, and secondly, in social networks. In the latter, we often give importance to the opinion of our friends, and the people we tend to follow. So in the end, a system which can automatically mine and extract our friends' opinions is likely to influence and benefit us.

8.2 Advantages

The proposed approach offers the following advantages in comparison to state-of-the-art:

- The proposed text sentiment analysis framework has been developed based on both concept and word level analysis. Concept level paradigm goes beyond mere word level analysis by focusing on the semantic relation between two words in a text. On the other hand, word level text analysis mainly focuses on ngrams. Combining concept and word level paradigms allow us to incorporate complementary information in the framework which shows performance improvement over state-of-the-art.

- The role of linguistics has mostly been ignored by state-of-the-art in sentiment analysis. Developing and employing linguistic patterns based on the syntactic relationship between words is a major benefit of the proposed approach. We show that these linguistic patterns are very useful and when combined with machine learning based approaches, the overall framework achieves impressive performance. By allowing sentiments to flow from concept to concept based on the dependency relation of the input sentence, we achieve a better understanding of the contextual role of each concept within the sentence and, hence, obtain a dynamic polarity detection engine that outperforms state-of-the-art statistical methods.
- Multimodal emotion recognition is a well known research topic since 1970. However, most of the works in this field were carried out using only audio or visual modality. Even in most recent studies on multimodal analysis, the complementary role of textual modality is ignored. In this work, we combine textual information with audio and visual information in order to improve multimodal sentiment analysis and emotion recognition. Comparative experimental results show that the use of textual modality adds key information to improve multimodal sentiment analysis. To this end, the developed framework was also applied to multimodal emotion recognition and obtained state-of-the-art performance.

8.3 Limitations

The proposed approach has the following limitations:

- Some of the techniques presented in this book, such as sentic patterns, work only at the sentence level and are not trivially generalized to document level. For example, a generalization of sentic patters to document level would involve discourse analysis instead of syntactic parsing.
- The sentic patterns technique can be applied only to grammatically correct sentences, since it involves a parsing stage. It also needs accurate and effective pre-processing to work well.
- The proposed methods for multi-modal sentiment analysis have been tested only on a limited dataset.
- In this research, we did not address audio feature extraction; instead, we just relied on an existing audio feature extraction system.
- Some of the techniques introduced in this book, such as concept extraction and sentic patterns, are language-dependent. We have developed specific rules only for English. Developing similar rules for other languages was not addressed in this book.
- Deep neural networks, on which some of our proposed techniques rely, are computationally expensive.

8.4 Future Work

For the future, we have planned more extensive evaluation. Also, real time validation is required using additional large datasets, compared against other state-of-the-art feature selection and classification approaches. Our other planned future endeavors are discussed below.

8.4.1 Brain Inspired Modeling of Sentiment

In order for machines to exploit common-sense knowledge in reasoning, as humans do, we need to endow them with human-like reasoning strategies as well. This can exploit the benefits of both evolutionary strategies and improved models of human performance. For example, psychological studies have shown that humans have evolved to use reasoning strategies that are often neither complete nor optimal, yet superior to pure logical reasoning.

When concepts become activated during reasoning, the spread of activation from one concept to another depends on how those concepts are connected. Such differences significantly alter perceptions, leading to decisive changes in the outcome of decisions or the activation of particular emotions or cognitive frames. For this reason, we plan to extend the two-level affective common-sense reasoning framework, as proposed in [43], by further investigating the impact of the presence and connectedness of knowledge base concepts of both conscious and unconscious cognition and reasoning. This could lead to the development of a brain-inspired sentiment analysis engine.

8.4.2 Aspect Level Sentiment Analysis

In the field of sentiment analysis, product features are termed aspects. They are mainly categorized into two categories – Implicit and Explicit aspects. Identifying aspects and detecting their polarity are well-known research problems in sentiment analysis. In this research we mainly dealt with them by extracting concepts. However, a more holistic future approach is required to extract all aspects in a document. Following the state of the art approach reported in [285], we plan to employ linguistic patterns and deep learning for extracting aspects and detecting polarity of aspects in the document. Although in the literature, many research works have been carried out on extracting aspects from opinionated documents, only a few researchers have focused on detecting polarity of those aspects. To solve this challenging problem, recent studies have proposed the use of Long Short Term Memory (LSTM) and Recursive Neural Networks (RNN). We plan to further extend those networks and integrate linguistic knowledge in deep networks to improve the aspect polarity assignment performance.

8.4.3 Document Level Sentiment Analysis

The proposed text-based sentiment analysis in this research was limited only to sentence level sentiment analysis, which was sufficient since we focused on utterance-based multimodal sentiment analysis. However, there are cases where we need to know the overall polarity of a document. This can be termed document polarity summarization. As a part of our future work, we plan to use deep learning based hierarchical Convolutional Neural Networks for document modeling. Such models will combine information extracted from every sentence present in a document for document modeling as shown in Fig. 8.1.

8.4.4 Multilingual Sentiment Analysis

The present framework can only process English. However, it is always desired that a sentiment analysis framework should be able to work well on any language [81]. Hence, we aim to extend our research towards multilingual analysis by employing domain adaptation and knowledge transfer techniques. Cross domain knowledge transfer approaches will be explored in the future, including, for example, by evaluating the performance of a English language trained system on Spanish language.

8.4.5 Detecting Sarcasm

Sarcasm detection is a key task [221, 287] for many natural language processing tasks. In sentiment analysis, for example, sarcasm can flip the polarity of an "apparently positive" sentence and, hence, negatively affect polarity detection performance. To date, most approaches to sarcasm detection have treated the task primarily as a text categorization problem. Sarcasm, however, can be expressed in very subtle ways and requires a deeper understanding of natural language that standard text categorization techniques cannot grasp. So in order to attack this research problem we aim to improve and employ our proposed text-based sentiment analysis engine which merges linguistic knowledge with machine learning.

8.4.6 Generalizing the Model

A real time system should work on any kind of real world data. In particular, an efficient system should not be biased towards a particular data type. This problem is often called as generalizability issue in the field of machine learning. The proposed

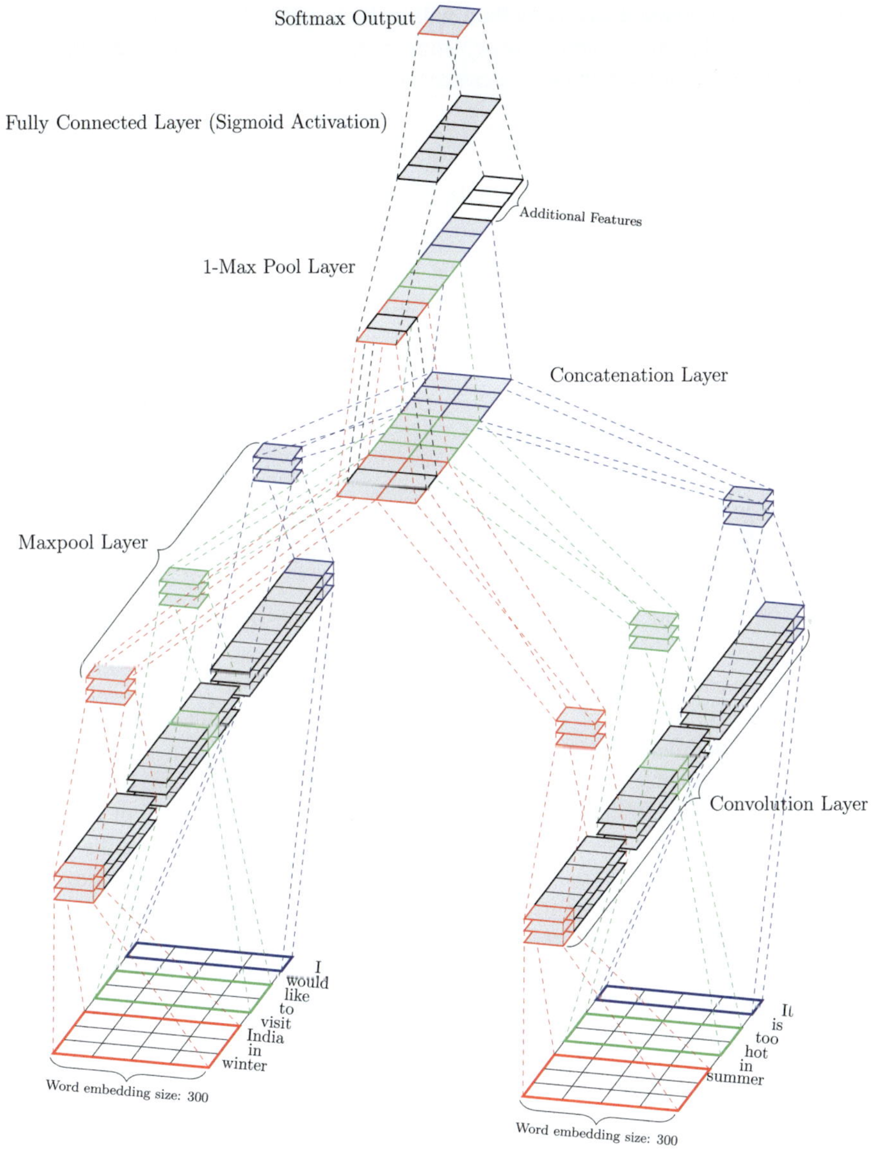

Fig. 8.1 The planned document modeling approach for sentiment analysis

multimodal sentiment analysis method will be built and validated on large datasets and across different domains. Apart from that, we also aim to understand the sentiment and emotion dynamics in conversations.a

8.4.7 Design Loss Function and Convergence Analysis

Designing task dependent loss functions help to customize network structure to achieve better result. As a part of our future work, we aim to design new loss functions based on the given task. As regular convergence and continuity checking is a rudimentary step in this research field, it is also an important aspect of our future work. We plan to use the research works by Rosasco et al. [325], Zhang et al. [425] as theoretical bases to start our analysis.

8.4.8 Use of Text Data in Social Computing

As the popularity of social network has grown dramatically in the last decade, it has become a new trend to analyze the influence of nodes in a social network. These nodes can be a person, group, or community etc. Most of the approaches to influence analysis are graph mining based methods. We argue that textual data can also be very helpful to measure influence of a node on another in social network. For example, suppose person A watched a movie and gave negative review. This review can 'negatively' influence person B, who is a friend or follower of person A, who is going to see the movie. Text analysis will help us to understand the review given by person A which cannot be analyzed using graph mining techniques. On the other hand, if most of the friends of person B gave positive reviews to the same movie, then that may nullify the influence of person A. Graph analysis will play a major role to mine such information. So, in the end we will build a hybrid system by integrating text analysis and social graph mining techniques for enhanced social network influence analysis.

References

1. Adams WH, Iyengar G, Lin C-Y, Naphade MR, Neti C, Nock HJ, Smith JR (2003) Semantic indexing of multimedia content using visual, audio, and text cues. EURASIP J Adv Signal Process 2003(2):1–16
2. Alam F, Riccardi G (2014) Predicting personality traits using multimodal information. In: Proceedings of the 2014 ACM Multimedia on Workshop on Computational Personality Recognition. ACM, pp 15–18
3. Alm CO, Roth D, Sproat R (2005) Emotions from text: machine learning for text-based emotion prediction. In: Proceedings of the Conference on Human Language Technology and Empirical Methods in Natural Language Processing. Association for Computational Linguistics, pp 579–586
4. Anand N, Verma P (2015) Convoluted feelings convolutional and recurrent nets for detecting emotion from audio data. Technical report, Stanford University
5. Anscombre J, Ducrot O (1977) Deux mais en français. Lingua 43:23–40
6. Arora P, Bakliwal A, Varma V (2012) Hindi subjective lexicon generation using wordnet graph traversal. Int J Comput Ling Appl 3(1):25–39
7. Asher N, Lascarides A (2003) Logics of conversation. Cambridge University Press, Cambridge
8. Asher N, Benamara F, Mathieu YY (2009) Appraisal of opinion expressions in discourse. Lingvisticæ Investigationes 32(2):279–292
9. Atassi H, Esposito A (2008) A speaker independent approach to the classification of emotional vocal expressions. In: 20th IEEE International Conference on Tools with Artificial Intelligence, ICTAI'08, vol 2. IEEE, pp 147–152
10. Atrey PK, Hossain MA, El Saddik A, Kankanhalli MS (2010) Multimodal fusion for multimedia analysis: a survey. Multimedia Syst 16(6):345–379
11. Awad M, Khan L, Bastani F, Yen I-L (2004) An effective support vector machines (svms) performance using hierarchical clustering. In: 16th IEEE International Conference on Tools with Artificial Intelligence, ICTAI'04. IEEE, pp 663–667
12. Bailenson JN, Pontikakis ED, Mauss IB, Gross JJ, Jabon ME, Hutcherson CAC, Nass C, John O (2008) Real-time classification of evoked emotions using facial feature tracking and physiological responses. Int J Hum Comput Stud 66(5):303–317
13. Bajpai R, Poria S, Ho D, Cambria E (2017) Developing a concept-level knowledge base for sentiment analysis in Singlish. In: CICLing, 2017
14. Balomenos T, Raouzaiou A, Ioannou S, Drosopoulos A, Karpouzis K, Kollias S (2004) Emotion analysis in man-machine interaction systems. In: Machine learning for multimodal interaction. Springer, pp 318–328

S. Poria et al., *Multimodal Sentiment Analysis*, Socio-Affective Computing 8,
https://doi.org/10.1007/978-3-319-95020-4

15. Baltrusaitis T, Banda N, Robinson P (2013) Dimensional affect recognition using continuous conditional random fields. In: 2013 10th IEEE International Conference and Workshops on Automatic Face and Gesture Recognition (FG). IEEE, pp 1–8
16. Baltrusaitis T, Robinson P, Morency L-P (2012) 3D constrained local model for rigid and non-rigid facial tracking. In: 2012 IEEE Conference on Computer Vision and Pattern Recognition (CVPR). IEEE, pp 2610–2617
17. See Ref. [16].
18. Banda N, Robinson P (2011) Noise analysis in audio-visual emotion recognition. In: International Conference on Multimodal Interaction, Alicante. Citeseer, pp 1–4
19. Bartlett MS, Hager JC, Ekman P, Sejnowski TJ (1999) Measuring facial expressions by computer image analysis. Psychophysiology 36(02):253–263
20. Batliner A, Fischer K, Huber R, Spilker J, Nöth E (2003) How to find trouble in communication. Speech Commun 40(1):117–143
21. Beaver D (2008) Presupposition and assertion in dynamic semantics. CSLI Publications
22. Bezdek JC (1981) Pattern recognition with fuzzy objective function algorithms. Kluwer Academic Publishers
23. Black MJ, Yacoob Y (1995) Tracking and recognizing rigid and non-rigid facial motions using local parametric models of image motion. In: Proceedings of the Fifth International Conference on Computer Vision, 1995. IEEE, pp 374–381
24. Blanz V, Vetter T (1999) A morphable model for the synthesis of 3D faces. In: Proceedings of the 26th Annual Conference on Computer Graphics and Interactive Techniques. ACM Press/Addison-Wesley Publishing Co, pp 187–194
25. See Ref. [26].
26. Blitzer J, Dredze M, Pereira F et al (2007) Biographies, Bollywood, boom-boxes and blenders: domain adaptation for sentiment classification. In: ACL, vol 7, pp 440–447
27. Boley D, Cao D (2004) Training support vector machines using adaptive clustering. In: SDM. SIAM, pp 126–137
28. Bollegala D, Weir D, Carroll J (2013) Cross-domain sentiment classification using a sentiment sensitive thesaurus. IEEE Trans Knowl Data Eng 25(8):1719–1731
29. Bravo-Marquez F, Mendoza M, Poblete B (2014) Meta-level sentiment models for big social data analysis. Knowl-Based Syst 69:86–99
30. Breidt M, Wallraven C, Cunningham DW, Bulthoff HH (2003) Facial animation based on 3D scans and motion capture. In: Siggraph'03 Sketches and Applications
31. Bughin J, Doogan J, Vetvik OJ (2010) A new way to measure word-of-mouth marketing. McKinsey Q 2:113–116
32. Burkhardt F, Paeschke A, Rolfes M, Sendlmeier WF, Weiss B (2005) A database of German emotional speech. In: Interspeech, vol 5, pp 1517–1520
33. Busso C, Bulut M, Lee C-C, Kazemzadeh A, Mower E, Kim S, Chang JN, Lee S, Narayanan SS (2008) Iemocap: interactive emotional dyadic motion capture database. Lang Resour Eval 42(4):335–359
34. Busso C, Deng Z, Yildirim S, Bulut M, Lee CM, Kazemzadeh A, Lee S, Neumann U, Narayanan S (2004) Analysis of emotion recognition using facial expressions, speech and multimodal information. In: Proceedings of the 6th International Conference on Multimodal Interfaces. ACM, pp 205–211
35. Cai G, Xia B (2015) Convolutional neural networks for multimedia sentiment analysis. In: National CCF Conference on Natural Language Processing and Chinese Computing, pp 159–167. Springer
36. Cambria E (2016) Affective computing and sentiment analysis. IEEE Intell Syst 31(2): 102–107
37. Cambria E, Fu J, Bisio F, Poria S (2015) AffectiveSpace 2: enabling affective intuition for concept-level sentiment analysis. In: AAAI, pp 508–514, Austin
38. See Ref. [37].
39. Cambria E, Gastaldo P, Bisio F, Zunino R (2015) An ELM-based model for affective analogical reasoning. Neurocomputing 149:443–455

40. Cambria E, Hazarika D, Poria S, Hussain A, Subramaanyam RBV (2017) Benchmarking multimodal sentiment analysis. arXiv preprint arXiv:1707.09538
41. Cambria E, Howard N, Hsu J, Hussain A (2013) Sentic blending: scalable multimodal fusion for continuous interpretation of semantics and sentics. In: IEEE SSCI, Singapore, pp 108–117
42. Cambria E, Hussain A (2015) Sentic computing: a common-sense-based framework for concept-level sentiment analysis, vol 1. Springer
43. Cambria E, Hussain A, Havasi C, Eckl C (2010) Senticspace: visualizing opinions and sentiments in a multi-dimensional vector space. In: Jordanov I, Setchi R (eds) Knowledge-based and intelligent information and engineering systems. Springer, Berlin/Heidelberg, pp 385–393
44. Cambria E, Livingstone A, Hussain A (2012) The hourglass of emotions. In: Esposito A, Vinciarelli A, Hoffmann R, Muller V (eds) Cognitive behavioral systems. Lecture notes in computer science, vol 7403. Springer, Berlin/Heidelberg, pp 144–157
45. See Ref. [46].
46. Cambria E, Poria S, Bajpai R, Schuller B (2016) Senticnet 4: a semantic resource for sentiment analysis based on conceptual primitives. In: Proceedings of the COLING 2016, the 26th International Conference on Computational Linguistics: Technical Papers, pp 2666–2677
47. Cambria E, Poria S, Bisio F, Bajpai R, Chaturvedi I (2015) The CLSA model: a novel framework for concept-level sentiment analysis. In: LNCS, vol 9042. Springer, pp 3–22
48. Cambria E, Poria S, Gelbukh A, Thelwall M (2017) Sentiment analysis is a big suitcase. IEEE Intell Syst 32(6):74–80
49. Cambria E, Poria S, Hazarika D, Kwok K (2018) Senticnet 5: discovering conceptual primitives for sentiment analysis by means of context embeddings. In: AAAI
50. Cambria E, White B (2014) Jumping NLP curves: a review of natural language processing research. IEEE Comput Intell Mag 9(2):48–57
51. Caridakis G, Castellano G, Kessous L, Raouzaiou A, Malatesta L, Asteriadis S, Karpouzis K (2007) Multimodal emotion recognition from expressive faces, body gestures and speech. In: Artificial intelligence and innovations 2007: from theory to applications. Springer, New York, pp 375–388
52. Caridakis G, Malatesta L, Kessous L, Amir N, Raouzaiou A, Karpouzis K (2006) Modeling naturalistic affective states via facial and vocal expressions recognition. In: Proceedings of the 8th International Conference on Multimodal Interfaces. ACM, pp 146–154
53. Castellano G, Kessous L, Caridakis G (2008) Emotion recognition through multiple modalities: face, body gesture, speech. In: Peter C, Beale R (eds) Affect and emotion in human-computer interaction. Springer, Berlin/Heidelberg, pp 92–103
54. Cervantes J, Li X, Yu W (2006) Support vector machine classification based on fuzzy clustering for large data sets. In: MICAI 2006: advances in artificial intelligence. Springer, Berlin/Heidelberg, pp 572–582
55. Chanel G, Rebetez C, Bétrancourt M, Pun T (2011) Emotion assessment from physiological signals for adaptation of game difficulty. IEEE Trans Syst Man Cybern Syst Hum 41(6): 1052–1063
56. Chang K-H, Fisher D, Canny J (2011) Ammon: a speech analysis library for analyzing affect, stress, and mental health on mobile phones. In: Proceedings of PhoneSense
57. Chang Y, Hu C, Turk M (2004) Probabilistic expression analysis on manifolds. In: Computer Vision and Pattern Recognition, vol 2. IEEE, pp II–520
58. Chaturvedi I, Ong Y-S, Tsang I, Welsch R, Cambria E (2016) Learning word dependencies in text by means of a deep recurrent belief network. Knowl-Based Syst 108:144–154
59. Chaturvedi I, Poria S, Cambria E (2017) Basic tasks of sentiment analysis. arXiv preprint arXiv:1710.06536
60. Chaumartin F-R (2007) Upar7: a knowledge-based system for headline sentiment tagging. In: Proceedings of the 4th International Workshop on Semantic Evaluations. Association for Computational Linguistics, pp 422–425.

61. Chawla K, Ramteke A (2013) Iitb-sentiment-analysts: participation in sentiment analysis in twitter semeval 2013 task. In: Second Joint Conference on Lexical and Computational Semantics (* SEM), vol 2. Citeseer, pp 495–500
62. Chen C-Y, Huang Y-K, Cook P (2005) Visual/acoustic emotion recognition. In: 2005 IEEE International Conference on Multimedia and Expo. IEEE, pp 1468–1471
63. Chen LS-H (2000) Joint processing of audio-visual information for the recognition of emotional expressions in human-computer interaction. Ph.D. thesis, Citeseer
64. Chen L, Yoon S-Y, Leong CW, Martin M, Ma M (2014) An initial analysis of structured video interviews by using multimodal emotion detection. In: Proceedings of the 2014 Workshop on Emotion Representation and Modelling in Human-Computer-Interaction-Systems. ACM, pp 1–6
65. Chenlo JM, Losada DE (2014) An empirical study of sentence features for subjectivity and polarity classification. Inform Sci 280:275–288
66. Chetty G, Wagner M, Goecke R (2008) A multilevel fusion approach for audiovisual emotion recognition. In: AVSP, pp 115–120
67. Chuang Z-J, Wu C-H (2004) Multi-modal emotion recognition from speech and text. Comput Ling Chin Lang Process 9(2):45–62
68. Cochrane T (2009) Eight dimensions for the emotions. Soc Sci Inf 48(3):379–420
69. Cohen I, Sebe N, Garg A, Chen LS, Huang TS (2003) Facial expression recognition from video sequences: temporal and static modeling. Comput Vis Image Underst 91(1):160–187
70. Cohen I, Sebe N, Gozman FG, Cirelo MC, Huang TS (2003) Learning Bayesian network classifiers for facial expression recognition both labeled and unlabeled data. In: 2003 IEEE Computer Society Conference on Computer Vision and Pattern Recognition Proceedings, vol 1. IEEE, pp I–595
71. Collobert R, Weston J, Bottou L, Karlen M, Kavukcuoglu K, Kuksa P (2011) Natural language processing (almost) from scratch. J Mach Learn Res 12:2493–2537
72. Cootes TF, Edwards GJ, Taylor CJ et al (2001) Active appearance models. IEEE Trans Pattern Anal Mach Intell 23(6):681–685
73. Cootes TF, Taylor CJ, Cooper DH, Graham J (1995) Active shape models-their training and application. Comput Vis Image Underst 61(1):38–59
74. Coppock E, Beaver D (2013) Principles of the exclusive muddle. J Semant. https://doi.org/10.1093/jos/fft007
75. Corradini A, Mehta M, Bernsen NO, Martin J, Abrilian S (2005) Multimodal input fusion in human-computer interaction. NATO Sci Ser Sub Ser III Comput Syst Sci 198:223
76. Cueva DR, Gonçalves RAM, Cozman F, Pereira-Barretto MR (2011) Crawling to improve multimodal emotion detection. In: Advances in soft computing. Springer, pp 343–350
77. Dalgleish T, Power MJ (1999) Handbook of cognition and emotion. Wiley Online Library, Chichester
78. Darwin C (1872) The expression of the emotions in man and animals. John Murray, London
79. Das D, Bandyopadhyay S (2011) Analyzing emotional statements–roles of general and physiological variables. In: The SAAIP Workshop, 5th IJCNLP. Citeseer, pp 59–67
80. Das D, Bandyopadhyay S (2012) Tracking emotions of bloggers–a case study for bengali. Polibits 45:53–59
81. Dashtipour K, Poria S, Hussain A, Cambria E, Hawalah AYA, Gelbukh A, Zhou Q (2016) Multilingual sentiment analysis: state of the art and independent comparison of techniques. Cogn Comput 8(4):757–771
82. Datcu D, Rothkrantz L (2008) Semantic audio-visual data fusion for automatic emotion recognition. In: Euromedia'2008
83. Datcu D, Rothkrantz LJM (2011) Emotion recognition using bimodal data fusion. In: Proceedings of the 12th International Conference on Computer Systems and Technologies. ACM, pp 122–128
84. Davatzikos C (2001) Measuring biological shape using geometry-based shape transformations. Image Vis Comput 19(1):63–74

85. De Meijer M (1989) The contribution of general features of body movement to the attribution of emotions. J Nonverbal Behav 13(4):247–268
86. Dellaert F, Polzin T, Waibel A (1996) Recognizing emotion in speech. In: Proceedings Fourth International Conference on Spoken Language ICSLP'96, vol 3. IEEE, pp 1970–1973
87. DeVault D, Artstein R, Benn G, Dey T, Fast E, Gainer A, Georgila K, Gratch J, Hartholt A, Lhommet M et al (2014) Simsensei kiosk: a virtual human interviewer for healthcare decision support. In: Proceedings of the 2014 International Conference on Autonomous Agents and Multi-Agent Systems, pp 1061–1068
88. Devillers L, Vidrascu L, Lamel L (2005) Challenges in real-life emotion annotation and machine learning based detection. Neural Netw 18(4):407–422
89. Ding X, Liu B, Yu PS (2008) A holistic lexicon-based approach to opinion mining. In: Proceedings of the 2008 International Conference on Web Search and Data Mining. ACM, pp 231–240
90. D'mello S, Graesser A (2007) Mind and body: dialogue and posture for affect detection in learning environments. Front Artif Intell Appl 158:161
91. D'Mello SK, Graesser A (2010) Multimodal semi-automated affect detection from conversational cues, gross body language, and facial features. User Model User-Adap Inter 20(2):147–187
92. D'mello SK, Kory J (2015) A review and meta analysis of multimodal affect detection systems. ACM Comput Surv 47(3):43–79
93. Dobrišek S, Gajšek R, Mihelič F, Pavešić N, Štruc V (2013) Towards efficient multi-modal emotion recognition. Int J Adv Robot Syst 10(53)
94. Donato G, Bartlett MS, Hager JC, Ekman P, Sejnowski TJ (1999) Classifying facial actions. IEEE Trans Pattern Anal Mach Intell 21(10):974–989
95. Douglas-Cowie E, Cowie R, Cox C, Amier N, Heylen DKJ (2008) The sensitive artificial listener: an induction technique for generating emotionally coloured conversation. In: LREC Workshop on Corpora for Research on Emotion and Affect
96. Douglas-Cowie E, Cowie R, Schroder M (2000) A new emotion database: considerations, sources and scope. In: ISCA Tutorial and Research Workshop (ITRW) on Speech and Emotion, pp 39–44
97. Douglas-Cowie E, Cowie R, Sneddon I, Cox C, Lowry O, Mcrorie M, Martin J-C, Devillers L, Abrilian S, Batliner A et al (2007) The humaine database: addressing the collection and annotation of naturalistic and induced emotional data. In: Affective computing and intelligent interaction. Springer, Berlin/Heidelberg, pp 488–500
98. Dragoni M, Poria S, Cambria E Ontosenticnet: a commonsense ontology for sentiment analysis
99. Dy MLIC, Espinosa IVL, Go PPV, Mendez CMM, Cu JW (2010) Multimodal emotion recognition using a spontaneous filipino emotion database. In: 2010 3rd International Conference on Human-Centric Computing (HumanCom). IEEE, pp 1–5
100. Egger M, Schoder D (2017) Consumer-oriented tech mining: integrating the consumer perspective into organizational technology intelligence – the case of autonomous driving. In: Hawaii International Conference on System Sciences
101. Ekenel HK, Sankur B (2004) Feature selection in the independent component subspace for face recognition. Pattern Recogn Lett 25(12):1377–1388
102. Ekman P, Rosenberg E, Hager J (1998) Facial action coding system affect interpretation dictionary (FACSAID)
103. Ekman P, Friesen WV, Hager JC (2002) Facs investigator's guide. A human face
104. Ekman P, Keltner D (1970) Universal facial expressions of emotion. Calif Mental Health Res Digest 8(4):151–158
105. Ekman P, Rosenberg EL (1997) What the face reveals: basic and applied studies of spontaneous expression using the facial action coding system (FACS). Oxford University Press
106. El Ayadi M, Kamel MS, Karray F (2011) Survey on speech emotion recognition: features, classification schemes, and databases. Pattern Recogn 44(3):572–587

107. Ellis JG, Jou B, Chang S-F (2014) Why we watch the news: a dataset for exploring sentiment in broadcast video news. In: Proceedings of the 16th International Conference on Multimodal Interaction. ACM, pp 104–111
108. Emerich S, Lupu E, Apatean A (2009) Emotions recognition by speech and facial expressions analysis. In: Proceedings of the 17th European Signal Processing Conference (EUSIPCO'09), pp 1617–1621
109. Esuli A, Sebastiani F (2006) Sentiwordnet: a publicly available lexical resource for opinion mining. In: Proceedings of LREC, vol 6. Citeseer, pp 417–422
110. Eyben F, Wöllmer M, Graves A, Schuller B, Douglas-Cowie E, Cowie R (2010) On-line emotion recognition in a 3-D activation-valence-time continuum using acoustic and linguistic cues. J Multimodal User Interfaces 3(1–2):7–19
111. Eyben F, Wöllmer M, Schuller B (2009) Openear—introducing the munich open-source emotion and affect recognition toolkit. In: 2009 3rd International Conference on Affective Computing and Intelligent Interaction and Workshops. IEEE, pp 1–6
112. Eyben F, Wöllmer M, Schuller B (2010) Opensmile: the munich versatile and fast open-source audio feature extractor. In: Proceedings of the 18th ACM International Conference on Multimedia. ACM, pp 1459–1462
113. Eyben F, Wöllmer M, Valstar MF, Gunes H, Schuller B, Pantic M (2011) String-based audiovisual fusion of behavioural events for the assessment of dimensional affect. In: 2011 IEEE International Conference on Automatic Face & Gesture Recognition and Workshops (FG 2011). IEEE, pp 322–329
114. Fasel B, Luettin J (2000) Recognition of asymmetric facial action unit activities and intensities. In: 15th International Conference on Pattern Recognition Proceedings (2000), vol 1. IEEE, pp 1100–1103
115. Fasel B, Luettin J (2003) Automatic facial expression analysis: a survey. Pattern Recogn 36(1):259–275
116. Filatova E (2012) Irony and sarcasm: corpus generation and analysis using crowdsourcing. In: LREC, pp 392–398
117. Fontaine JRJ, Scherer KR, Roesch EB, Ellsworth PC (2007) The world of emotions is not two-dimensional. Psychol Sci 18(12):1050–1057
118. Forbes-Riley K, Litman DJ (2004) Predicting emotion in spoken dialogue from multiple knowledge sources. In: HLT-NAACL. Citeseer, pp 201–208
119. Freitas A, Castro E (2009) Facial expression: the effect of the smile in the treatment of depression. Empirical study with Portuguese subjects. In: Emotional expression: the brain and the face, pp 127–140
120. Gajsek R, Štruc V, Mihelic F (2010) Multi-modal emotion recognition using canonical correlations and acoustic features. In: Proceedings of the 2010 20th International Conference on Pattern Recognition. IEEE Computer Society, pp 4133–4136
121. Gale WA, Church KW, Yarowsky D (1992) One sense per discourse. In: Proceedings of the Workshop on Speech and Natural Language. Association for Computational Linguistics, pp 233–237
122. Gangemi A, Presutti V, Reforgiato Recupero D (2014) Frame-based detection of opinion holders and topics: a model and a tool. IEEE Comput Intell Mag 9(1):20–30
123. Gezici G, Dehkharghani R, Yanikoglu B, Tapucu D, Saygin Y (2013) Su-sentilab: a classification system for sentiment analysis in twitter. In: International Workshop on Semantic Evaluation, pp 471–477
124. Glodek M, Reuter S, Schels M, Dietmayer K, Schwenker F (2013) Kalman filter based classifier fusion for affective state recognition. In: Multiple classifier systems, pp 85–94. Springer
125. Glodek M, Tschechne S, Layher G, Schels M, Brosch T, Scherer S, Kächele M, Schmidt M, Neumann H, Palm G et al (2011) Multiple classifier systems for the classification of audio-visual emotional states. In: Affective computing and intelligent interaction, pp 359–368. Springer

126. Glorot X, Bordes A, Bengio Y (2011) Domain adaptation for large-scale sentiment classification: a deep learning approach. In: Proceedings of the Twenty-Eight International Conference on Machine Learning, ICML, pp 513–520
127. Go A, Bhayani R, Huang L (2009) Twitter sentiment classification using distant supervision. CS224N Project Report, Stanford, pp 1–12
128. Gong S, Shan C, Xiang T (2007) Visual inference of human emotion and behaviour. In: Proceedings of the 9th International Conference on Multimodal Interfaces. ACM, pp 22–29
129. Graves A, Fernández S, Schmidhuber J (2005) Bidirectional LSTM networks for improved phoneme classification and recognition. In: Artificial neural networks: formal models and their applications–ICANN 2005. Springer, pp 799–804
130. Gunes H, Pantic M (2010) Dimensional emotion prediction from spontaneous head gestures for interaction with sensitive artificial listeners. In: International Conference on Intelligent Virtual Agents, pp 371–377
131. Gunes H, Piccardi M (2005) Affect recognition from face and body: early fusion vs. late fusion. In: 2005 IEEE International Conference on Systems, Man and Cybernetics, vol 4. IEEE, pp 3437–3443
132. Gunes H, Piccardi M (2005) Fusing face and body display for bi-modal emotion recognition: single frame analysis and multi-frame post integration. In: Affective computing and intelligent interaction. Springer, pp 102–111
133. Gunes H, Piccardi M (2009) Automatic temporal segment detection and affect recognition from face and body display. IEEE Trans Syst Man Cybern B Cybern 39(1):64–84
134. Hamel P, Eck D (2010) Learning features from music audio with deep belief networks. In: ISMIR, Utrecht, pp 339–344
135. Han K, Yu D, Tashev I (2014) Speech emotion recognition using deep neural network and extreme learning machine. In: Interspeech, pp 223–227
136. Han M-J, Hsu J-H, Song K T, Chang F-Y (2007) A new information fusion method for SVM-based robotic audio-visual emotion recognition. In: IEEE International Conference on Systems, Man and Cybernetics ISIC (2007). IEEE, pp 2656–2661
137. Hanan AHA, Idris MY, Kaiwartya O, Prasad M, Shah RR (2017) Real traffic data based evaluation of vehicular traffic environment and state-of-the-art with future issues in location centric data dissemination for VANETs. Elsevier Digit Commun Netw 3(3):195–210
138. Haq S, Jackson PJB, Edge J (2009) Speaker-dependent audio-visual emotion recognition. In: AVSP, pp 53–58
139. Haq S, Jackson PJB, Edge J (2008) Audio-visual feature selection and reduction for emotion classification. In: Proceedings of International Conference on Auditory-Visual Speech Processing (AVSP'08), Tangalooma
140. Haro A, Flickner M, Essa I (2000) Detecting and tracking eyes by using their physiological properties, dynamics, and appearance. In: Computer vision and pattern recognition, vol 1. IEEE, pp 163–168
141. Hatzivassiloglou V, McKeown KR (1997) Predicting the semantic orientation of adjectives. In: Proceedings of the 35th Annual Meeting of the Association for Computational Linguistics and Eighth Conference of the European Chapter of the Association for Computational Linguistics. Association for Computational Linguistics, pp 174–181
142. Hinton G (2010) A practical guide to training restricted Boltzmann machines. Momentum 9(1):926
143. Hinton GE, Osindero S, Teh Y-W (2006) A fast learning algorithm for deep belief nets. Neural Comput 18(7):1527–1554
144. Hirschberg J, Benus S, Brenier JM, Enos F, Friedman S, Gilman S, Girand C, Graciarena M, Kathol A, Michaelis L et al (2005) Distinguishing deceptive from non-deceptive speech. In: INTERSPEECH, pp 1833–1836
145. Hirst G, St-Onge D (1998) Lexical chains as representations of context for the detection and correction of malapropisms. In: Fellbaum C (ed) WordNet: an electronic lexical database, vol 305. MIT Press, Cambridge/London, pp 305–332

146. Hoch S, Althoff F, McGlaun G, Rigoll G (2005) Bimodal fusion of emotional data in an automotive environment. In: IEEE International Conference on Acoustics, Speech, and Signal Processing Proceedings (ICASSP'05), vol 2. IEEE, pp II–1085
147. Hommel S, Rabie A, Handmann U (2013) Attention and emotion based adaption of dialog systems. In: Intelligent systems: models and applications. Springer, Berlin, pp 215–235
148. Hoque ME, Picard RW (2011) Acted vs. natural frustration and delight: many people smile in natural frustration. In: Automatic face & gesture recognition and workshops. IEEE, pp 354–359
149. Hu M, Liu B (2004) Mining and summarizing customer reviews. In: Proceedings of the Tenth ACM SIGKDD International Conference on Knowledge Discovery and Data Mining. ACM, pp 168–177
150. Hu X, Tang J, Gao H, Liu H (2013) Unsupervised sentiment analysis with emotional signals. In: Proceedings of the 22nd International Conference on World Wide Web. International World Wide Web Conferences Steering Committee, pp 607–618
151. Huang G-B (2014) An insight into extreme learning machines: random neurons, random features and kernels. Cogn Comput. https://doi.org/10.1007/s12559-014-9255-2
152. Huang G-B, Cambria E, Toh K-A, Widrow B, Xu Z (2015) New trends of learning in computational intelligence. IEEE Comput Intell Mag 10(2):16–17
153. Huang G-B, Cambria E, Toh K-A, Widrow B, Xu Z (2015) New trends of learning in computational intelligence (part II). IEEE Comput Intell Mag 10(3):8
154. Huang G-B, Chen L, Siew C-K (2006) Universal approximation using incremental constructive feedforward networks with random hidden nodes. IEEE Trans Neural Netw 17(4): 879–892
155. Huang G-B, Wang DH, Lan Y (2011) Extreme learning machines: a survey. Int J Mach Learn Cybern 2(2):107–122
156. Huang G-B, Zhou H, Ding X, Zhang R (2012) Extreme learning machine for regression and multiclass classification. IEEE Trans Syst Man Cybern B Cybern 42(2):513–529
157. Huang Z, Dong M, Mao Q, Zhan Y (2014) Speech emotion recognition using CNN. In: Proceedings of the ACM International Conference on Multimedia. ACM, pp 801–804
158. Hussain MS, Monkaresi H, Calvo RA (2012) Combining classifiers in multimodal affect detection. In: Proceedings of the Tenth Australasian Data Mining Conference-Volume 134. Australian Computer Society, Inc., pp 103–108
159. Iyengar G, Nock HJ, Neti C (2003) Audio-visual synchrony for detection of monologues in video archives. In: 2003 International Conference on Multimedia and Expo ICME'03 (2003) Proceedings, vol 1. IEEE, pp I–329
160. Izard CE, Dougherty LM, Hembree EA (1983) A system for identifying affect expressions by holistic judgments (AFFEX). Instructional Resources Center, University of Delaware
161. Jain A, Vishwanathan SVN, Varma M (2012) SPF-GMKL: generalized multiple kernel learning with a million kernels. In: Proceedings of the 18th ACM SIGKDD International Conference on Knowledge Discovery and Data Mining. ACM, pp 750–758
162. Jaiswal M, Tabibu S, Bajpai R (2016) The truth and nothing but the truth: multimodal analysis for deception detection. In: ICDM
163. Jayez J, Winterstein G (2013) Additivity and probability. Lingua 132:85–102
164. Ji R, Cao D, Lin D (2015) Cross-modality sentiment analysis for social multimedia. In: 2015 IEEE International Conference on Multimedia Big Data (BigMM). IEEE, pp 28–31
165. Jia L, Yu C, Meng W (2009) The effect of negation on sentiment analysis and retrieval effectiveness. In: Proceedings of the 18th ACM Conference on Information and Knowledge Management. ACM, pp 1827–1830
166. Jiang D, Cui Y, Zhang X, Fan P, Ganzalez I, Sahli H (2011) Audio visual emotion recognition based on triple-stream dynamic Bayesian network models. In: D'Mello S (ed) Affective computing and intelligent interaction. Springer, Berlin, pp 609–618
167. Jiang JJ, Conrath DW (1997) Semantic similarity based on corpus statistics and lexical taxonomy. arXiv preprint cmp-lg/9709008

168. Jin Q, Li C, Chen S, Wu H (2015) Speech emotion recognition with acoustic and lexical features. In: 2015 IEEE International Conference on Acoustics, Speech and Signal Processing (ICASSP). IEEE, pp 4749–4753
169. Joachims T (1998) Text categorization with support vector machines: learning with many relevant features
170. Johnstone T (1996) Emotional speech elicited using computer games. In: Fourth International Conference on Spoken Language ICSLP 96 (1996) Proceedings, vol 3. IEEE, pp 1985–1988
171. Jolliffe I (2002) Principal component analysis. Wiley Online Library
172. Jones MJ, Poggio T (1998) Multidimensional morphable models. In: Sixth International Conference on Computer Vision. IEEE, pp 683–688
173. Jong-Tae J, Sang-Wook S, Kwang-Eun K, Kwee-Bo S (2007) Emotion recognition method based on multimodal sensor fusion algorithm. ISIS, Sokcho-City
174. Jordan A (2002) On discriminative vs. generative classifiers: a comparison of logistic regression and naive bayes. Adv Neural Inf Proces Syst 14:841
175. Kaiwartya O, Abdullah AH, Cao Y, Raw RS, Kumar S, Lobiyal DK, Isnin IF, Liu X, Shah RR (2016) T-MQM: testbed-based multi-metric quality measurement of sensor deployment for precision agriculture—a case study. IEEE Sens J 16(23):8649–8664
176. Kalchbrenner N, Grefenstette E, Blunsom P (2014) A convolutional neural network for modelling sentences. CoRR, abs/1404.2188
177. Kanayama H, Nasukawa T (2006) Fully automatic lexicon expansion for domain-oriented sentiment analysis. In: Proceedings of the 2006 Conference on Empirical Methods in Natural Language Processing. Association for Computational Linguistics, pp 355–363
178. Kanluan I, Grimm M, Kroschel K (2008) Audio-visual emotion recognition using an emotion space concept. In: 2008 16th European Signal Processing Conference. IEEE, pp 1–5
179. Kapoor A, Picard RW (2005) Multimodal affect recognition in learning environments. In: Proceedings of the 13th Annual ACM International Conference on Multimedia. ACM, pp 677–682
180. Kapur A, Kapur A, Virji-Babul N, Tzanetakis G, Driessen PF (2005) Gesture-based affective computing on motion capture data. In: Affective computing and intelligent interaction. Springer, Berlin, pp 1–7
181. Karpathy A, Toderici G, Shetty S, Leung T, Sukthankar R, Fei-Fei L (2014) Large-scale video classification with convolutional neural networks. In: CVPR
182. Karpouzis K, Caridakis G, Kessous L, Amir N, Raouzaiou A, Malatesta L, Kollias S (2007) Modeling naturalistic affective states via facial, vocal, and bodily expressions recognition. In: Artificial intelligence for human computing. Springer, Berlin/Heidelberg, pp 91–112
183. Karttunen L (1973) Presuppositions of compound sentences. Ling Inq 4(2):169–193
184. Kavukcuoglu K, Sermanet P, Boureau Y-L, Gregor K, Mathieu M, Cun YL (2010) Learning convolutional feature hierarchies for visual recognition. In: Advances in neural information processing systems, pp 1090–1098
185. Kessous L, Castellano G, Caridakis G (2010) Multimodal emotion recognition in speech-based interaction using facial expression, body gesture and acoustic analysis. J Multimodal User Interfaces 3(1–2):33–48
186. Khalili Z, Moradi MH (2009) Emotion recognition system using brain and peripheral signals: using correlation dimension to improve the results of EEG. In: 2009 International Joint Conference on Neural Networks. IEEE, pp 1571–1575
187. Kim J (2007) Bimodal emotion recognition using speech and physiological changes. INTECH Open Access Publisher
188. Kim J, Lingenfelser F (2010) Ensemble approaches to parametric decision fusion for bimodal emotion recognition. In: BIOSIGNALS, pp 460–463
189. Kim Y (2014) Convolutional neural networks for sentence classification. arXiv preprint arXiv:1408.5882
190. Kim Y-H, Hahn S-Y, Zhang B-T (2000) Text filtering by boosting naive Bayes classifiers. In: Proceedings of the 23rd Annual International ACM SIGIR Conference on Research and Development in Information Retrieval. ACM, pp 168–175

191. Kimura S, Yachida M (1997) Facial expression recognition and its degree estimation. In: 1997 IEEE Computer Society Conference on Computer Vision and Pattern Recognition Proceedings. IEEE, pp 295–300
192. Koelstra S, Muhl C, Soleymani M, Lee J-S, Yazdani A, Ebrahimi T, Pun T, Nijholt A, Patras I (2012) Deap: a database for emotion analysis; using physiological signals. IEEE Trans Affect Comput 3(1):18–31
193. Koolagudi SG, Kumar N, Sreenivasa Rao K (2011) Speech emotion recognition using segmental level prosodic analysis. In: 2011 International Conference on Devices and Communications (ICDeCom). IEEE, pp 1–5
194. Krell G, Glodek M, Panning A, Siegert I, Michaelis B, Wendemuth A, Schwenker F (2013) Fusion of fragmentary classifier decisions for affective state recognition. In: Multimodal pattern recognition of social signals in human-computer-interaction. Springer, pp 116–130
195. Kring AM, Sloan D (1991) The facial expression coding system (faces): a users guide. Unpublished manuscript
196. Kring AM, Sloan DM (2007) The facial expression coding system (faces): development, validation, and utility. Psychol Assess 19(2):210
197. Krizhevsky A, Sutskever I, Hinton GE (2012) Imagenet classification with deep convolutional neural networks. In: Advances in neural information processing systems, pp 1097–1105
198. Krizhevsky A, Sutskever I, Geoffrey E (2012) Hinton. Imagenet classification with deep convolutional neural networks. In: Pereira F, Burges CJC, Bottou L, Weinberger KQ (eds) Advances in neural information processing systems, vol 25. Curran Associates, Inc, pp 1097–1105
199. Lanitis A, Taylor CJ, Cootes TF (1995) Automatic face identification system using flexible appearance models. Image Vis Comput 13(5):393–401
200. Lau RYK, Xia Y, Ye Y (2014) A probabilistic generative model for mining cybercriminal networks from online social media. IEEE Comput Intell Mag 9(1):31–43
201. Leacock C, Chodorow M (1998) Combining local context and wordnet similarity for word sense identification. In: Fellbaum C (ed) WordNet: an electronic lexical database, vol 49, pp 265–283. MIT Press, Cambridge/London
202. LeCun Y, Kavukcuoglu K, Farabet C et al (2010) Convolutional networks and applications in vision. In: ISCAS, pp 253–256
203. Lee CM, Narayanan SS (2005) Toward detecting emotions in spoken dialogs. IEEE Trans Speech Audio Process 13(2):293–303
204. Levenson RW (1994) Human emotion: a functional view. In: Ekman P, Davidson RJ (eds) The nature of emotion: fundamental questions, vol 1. Oxford University Press, New York, pp 123–126
205. Li Y, McLean D, Bandar ZA, O'shea JD, Crockett K (2006) Sentence similarity based on semantic nets and corpus statistics. IEEE Trans Knowl Data Eng 18(8):1138–1150
206. Lien JJ-J, Kanade T, Cohn JF, Li C-C (2000) Detection, tracking, and classification of action units in facial expression. Robot Auton Syst 31(3):131–146
207. Lin D (1998) An information-theoretic definition of similarity. In: ICML, vol 98, pp 296–304
208. Lin J-C, Wu C-H, Wei W-L (2012) Error weighted semi-coupled hidden Markov model for audio-visual emotion recognition. IEEE Trans Multimedia 14(1):142–156
209. Lin KH-Y, Yang C, Chen H-H (2007) What emotions do news articles trigger in their readers? In: Proceedings of the 30th Annual International ACM SIGIR Conference on Research and Development in Information Retrieval. ACM, pp 733–734
210. Lin Z, Hwee T, Kan M-Y (2012) A PDTB-styled end-to-end discourse parser. Nat Lang Eng
211. Lingenfelser F, Wagner J, André E (2011) A systematic discussion of fusion techniques for multi-modal affect recognition tasks. In: Proceedings of the 13th International Conference on Multimodal Interfaces. ACM, pp 19–26
212. Litman DJ, Forbes-Riley K (2004) Predicting student emotions in computer-human tutoring dialogues. In: Proceedings of the 42nd Annual Meeting on Association for Computational Linguistics. Association for Computational Linguistics, p 351

213. Litman DJ, Forbes-Riley K (2006) Recognizing student emotions and attitudes on the basis of utterances in spoken tutoring dialogues with both human and computer tutors. Speech Commun 48(5):559–590
214. Littlewort G, Bartlett MS, Fasel I, Susskind J, Movellan J (2006) Dynamics of facial expression extracted automatically from video. Image Vis Comput 24(6):615–625
215. Liu B (2012) Sentiment analysis and opinion mining. Synth Lect Hum Lang Technol 5(1): 1–167
216. Liu X, Wang L, Huang G-B, Zhang J, Yin J (2015) Multiple kernel extreme learning machine. Neurocomputing 149(Part A):253–264
217. Lu K, Jia Y (2012) Audio-visual emotion recognition with boosted coupled HMM. In: 2012 21st International Conference on Pattern Recognition (ICPR). IEEE, pp 1148–1151
218. Luengo I, Navas E, Hernáez I, Sánchez J (2005) Automatic emotion recognition using prosodic parameters. In: Interspeech, pp 493–496
219. Lyons MJ, Budynek J, Akamatsu S (1999) Automatic classification of single facial images. IEEE Trans Pattern Anal Mach Intell 21(12):1357–1362
220. Mairesse F, Walker MA, Mehl MR, Moore RK (2007) Using linguistic cues for the automatic recognition of personality in conversation and text. J Artif Intell Res
221. Majumder N, Poria S, Gelbukh A, Cambria E (2017) Deep learning-based document modeling for personality detection from text. IEEE Intell Syst 32(2):74–79
222. Mansoorizadeh M, Charkari NM (2010) Multimodal information fusion application to human emotion recognition from face and speech. Multimedia Tools Appl 49(2):277–297
223. Martin O, Kotsia I, Macq B, Pitas I (2006) The enterface'05 audio-visual emotion database. In: 22nd International Conference on Data Engineering Workshops Proceedings. IEEE, pp 8–8
224. Matsumoto D (1992) More evidence for the universality of a contempt expression. Motiv Emot 16(4):363–368
225. Matthews G, Gilliland K (1999) The personality theories of HJ Eysenck and JA Gray: a comparative review. Personal Individ Differ 26(4):583–626
226. McKeown G, Valstar M, Cowie R, Pantic M, Schroder M (2012) The semainc database: annotated multimodal records of emotionally colored conversations between a person and a limited agent. IEEE Trans Affect Comput 3(1):5–17
227. Mehrabian A (1996) Pleasure-arousal-dominance: a general framework for describing and measuring individual differences in temperament. Curr Psychol 14(4):261–292
228. Melville P, Gryc W, Lawrence RD (2009) Sentiment analysis of blogs by combining lexical knowledge with text classification. In: Proceedings of the 15th ACM SIGKDD International Conference on Knowledge Discovery and Data mining. ACM, pp 1275–1284
229. Metallinou A, Lee S, Narayanan S (2008) Audio-visual emotion recognition using gaussian mixture models for face and voice. In: Tenth IEEE International Symposium on Multimedia ISM 2008. IEEE, pp 250–257
230. Metallinou A, Wöllmer M, Katsamanis A, Eyben F, Schuller B, Narayanan S (2012) Context-sensitive learning for enhanced audiovisual emotion classification. IEEE Trans Affect Comput 3(2):184–198
231. See Ref. [232].
232. Mikolov T, Chen K, Corrado G, Dean J (2013) Efficient estimation of word representations in vector space. CoRR, abs/1301.3781
233. Mikolov T, Sutskever I, Chen K, Gregory S (2013) Corrado, Jeffrey Dean. Distributed representations of words and phrases and their compositionality. In: Advances in Neural Information Processing Systems 26: 27th Annual Conference on Neural Information Processing Systems 2013. Proceedings of a Meeting Held 5–8 Dec 2013, Lake Tahoe, Nevada, pp 3111–3119
234. Miller GA (1995) Wordnet: a lexical database for English. Commun ACM 38(11):39–41
235. Mishne G et al (2005) Experiments with mood classification in blog posts. In: Proceedings of ACM SIGIR 2005 Workshop on Stylistic Analysis of Text for Information Access, vol 19. Citeseer, pp 321–327

236. Mohammad SM, Kiritchenko S (2012) Using nuances of emotion to identify personality. AAAI Technical Report WS-13-01. In: Computational personality recognition (Shared Task)
237. Moilanen K, Pulman S (2007) Sentiment composition. In: Proceedings of the Recent Advances in Natural Language Processing International Conference, pp 378–382
238. Monkaresi H, Sazzad Hussain M, Calvo RA (2012) Classification of affects using head movement, skin color features and physiological signals. In: 2012 IEEE International Conference on Systems, Man, and Cybernetics (SMC). IEEE, pp 2664–2669
239. Morency L-P, Mihalcea R, Doshi P (2011) Towards multimodal sentiment analysis: harvesting opinions from the web. In: Proceedings of the 13th International Conference on Multimodal Interfaces. ACM, pp 169–176
240. Morency L-P, Whitehill J, Movellan J (2008) Generalized adaptive view-based appearance model: integrated framework for monocular head pose estimation. In: 8th IEEE International Conference on Automatic Face & Gesture Recognition FG'08. IEEE, pp 1–8
241. Morrison D, Wang R, De Silva LC (2007) Ensemble methods for spoken emotion recognition in call-centres. Speech Commun 49(2):98–112
242. Murray IR, Arnott JL (1993) Toward the simulation of emotion in synthetic speech: a review of the literature on human vocal emotion. J Acoust Soc Am 93(2):1097–1108
243. Narayanan R, Liu B, Choudhary A (2009) Sentiment analysis of conditional sentences. In: Proceedings of the 2009 Conference on Empirical Methods in Natural Language Processing, vol 1. Association for Computational Linguistics, pp 180–189
244. Navas E, Hernaez I, Luengo I (2006) An objective and subjective study of the role of semantics and prosodic features in building corpora for emotional TTS. IEEE Trans Audio Speech Lang Process 14(4):1117–1127
245. Nefian AV, Liang L, Pi X, Liu X, Murphy K (2002) Dynamic Bayesian networks for audio-visual speech recognition. EURASIP J Adv Signal Process 2002(11):1–15
246. Ng HT, Goh WB, Low KL (1997) Feature selection, perceptron learning, and a usability case study for text categorization. In: ACM SIGIR Forum, vol 31. ACM, pp 67–73
247. Nickel K, Gehrig T, Stiefelhagen R, McDonough J (2005) A joint particle filter for audio-visual speaker tracking. In: Proceedings of the 7th International Conference on Multimodal Interfaces. ACM, pp 61–68
248. Nicolaou MA, Gunes H, Pantic M (2010) Automatic segmentation of spontaneous data using dimensional labels from multiple coders. In: Proceedings of LREC International Workshop on Multimodal Corpora: Advances in Capturing, Coding and Analyzing Multimodality, pp 43–48
249. Nicolaou MA, Gunes H, Pantic M (2011) Continuous prediction of spontaneous affect from multiple cues and modalities in valence-arousal space. IEEE Trans Affect Comput 2(2): 92–105
250. Ohta H, Saji H, Nakatani H (1998) Recognition of facial expressions using muscle-based feature models. In: Fourteenth International Conference on Pattern Recognition Proceedings, vol 2. IEEE, pp 1379–1381
251. Oneto L, Bisio F, Cambria E, Anguita D (2016) Statistical learning theory and ELM for big social data analysis. IEEE Comput Intell Mag 11(3):45–55
252. Ortony A, Turner TJ (1990) What's basic about basic emotions? Psychol Rev 97(3):315
253. Osgood CE (1952) The nature and measurement of meaning. Psychol Bull 49(3):197–237
254. Pak A, Paroubek P (2010) Twitter as a corpus for sentiment analysis and opinion mining. In: LREc, vol 10, pp 1320–1326
255. Pakray P, Neogi S, Bhaskar P, Poria S, Bandyopadhyay S, Gelbukh AF (2011) A textual entailment system using anaphora resolution. In: TAC
256. Pakray P, Pal S, Poria S, Bandyopadhyay S, Gelbukh AF (2010) Ju_cse_tac: textual entailment recognition system at tac rte-6. In: TAC
257. Pakray P, Poria S, Bandyopadhyay S, Gelbukh A (2011) Semantic textual entailment recognition using UNL. Polibits 43:23–27
258. Pal P, Iyer AN, Yantorno RE (2006) Emotion detection from infant facial expressions and cries. In: 2006 IEEE International Conference on Acoustics, Speech and Signal Processing ICASSP 2006 Proceedings, vol 2. IEEE, pp II–II

259. Paleari M, Benmokhtar R, Huet B (2009) Evidence theory-based multimodal emotion recognition. In: International Conference on Multimedia Modeling. Springer, pp 435–446
260. Pan SJ, Ni X, Sun J-T, Yang Q, Chen Z (2010) Cross-domain sentiment classification via spectral feature alignment. In: Proceedings of the 19th International Conference on World Wide Web. ACM, pp 751–760
261. Pang B, Lee L (2005) Seeing stars: exploiting class relationships for sentiment categorization with respect to rating scales. In: ACL, Ann Arbor, pp 115–124
262. Pang B, Lee L, Vaithyanathan S (2002) Thumbs up?: sentiment classification using machine learning techniques. In: Proceedings of the ACL-02 Conference on Empirical Methods in Natural Language Processing, vol 10. Association for Computational Linguistics, pp 79–86
263. Pantic M, Rothkrantz LJM (2000) Automatic analysis of facial expressions: the state of the art. IEEE Trans Pattern Anal Mach Intell 22(12):1424–1445
264. Pantic M, Rothkrantz LJM (2000) Expert system for automatic analysis of facial expressions. Image Vis Comput 18(11):881–905
265. Park J-S, Jang G-J, Seo Y-H (2012) Music-aided affective interaction between human and service robot. EURASIP J Audio Speech Music Process 2012(1):1–13
266. Parke FI, Waters K (2008) Computer facial animation. CRC Press
267. Parrott WG (2001) Emotions in social psychology: essential readings. Psychology Press, Philadelphia
268. Patwardhan S, Banerjee S, Pedersen T (2003) Using measures of semantic relatedness for word sense disambiguation. In: Computational linguistics and intelligent text processing, pp 241–257
269. Paul E, Friesen W (1978) Facial action coding system investigator's guide. Consulting Psychologists Press, Palo Alto
270. Pedersen T, Patwardhan S, Michelizzi J (2004) Wordnet: Similarity: measuring the relatedness of concepts. In: Demonstration Papers at HLT-NAACL 2004. Association for Computational Linguistics, pp 38–41
271. Peersman C, Daelemans W, Van Vaerenbergh L (2011) Predicting age and gender in online social networks. In: Proceedings of the 3rd International Workshop on Search and Mining User-Generated Contents. ACM, pp 37–44
272. Pennington J, Socher R, Manning CD (2014) Glove: global vectors for word representation. In: EMNLP, vol 14, pp 1532–43
273. Pérez-Rosas V, Mihalcea R, Morency L-P (2013) Utterance-level multimodal sentiment analysis. In: ACL, vol 1, pp 973–982
274. Petridis S, Pantic M (2008) Audiovisual discrimination between laughter and speech. In: IEEE International Conference on Acoustics, Speech and Signal Processing ICASSP 2008. IEEE, pp 5117–5120
275. Phan X-H, Nguyen L-M, Horiguchi S (2008) Learning to classify short and sparse text & web with hidden topics from large-scale data collections. In: Proceedings of the 17th International Conference on World Wide Web. ACM, pp 91–100
276. Piana S, Staglianò A, Camurri A, Odone F (2013) A set of full-body movement features for emotion recognition to help children affected by autism spectrum condition. In: IDGEI International Workshop
277. Piana S, Stagliano A, Odone F, Verri A, Camurri A (2014) Real-time automatic emotion recognition from body gestures. arXiv preprint arXiv:1402.5047
278. Plutchik R (1980) Emotion: a psychoevolutionary synthesis. Harpercollins College Division
279. Plutchik R (2001) The nature of emotions human emotions have deep evolutionary roots, a fact that may explain their complexity and provide tools for clinical practice. Am Sci 89(4):344–350
280. Polanyi L, Culy C, Van Den Berg M, Thione GL, Ahn D (2004) Sentential structure and discourse parsing. In: Proceedings of the 2004 ACL Workshop on Discourse Annotation. Association for Computational Linguistics, pp 80–87

281. Poria S (2017) Novel symbolic and machine-learning approaches for text-based and multimodal sentiment analysis
282. Poria S, Cambria E, Bajpai R, Hussain A (2017) A review of affective computing: from unimodal analysis to multimodal fusion. Inf Fusion 37:98–125
283. Poria S, Cambria E, Gelbukh A (2015) Deep convolutional neural network textual features and multiple kernel learning for utterance-level multimodal sentiment analysis. In: Proceedings of the 2015 Conference on Empirical Methods in Natural Language Processing, pp 2539–2544
284. Poria S, Cambria E, Gelbukh A (2016) Aspect extraction for opinion mining with a deep convolutional neural network. Knowl-Based Syst 108:42–49
285. See Ref. [283].
286. Poria S, Cambria E, Gelbukh A, Bisio F, Hussain A (2015) Sentiment data flow analysis by means of dynamic linguistic patterns. IEEE Comput Intell Mag 10(4):26–36
287. Poria S, Cambria E, Hazarika D, Vij P (2016) A deeper look into sarcastic tweets using deep convolutional neural networks. In: COLING, pp 1601–1612
288. Poria S, Cambria E, Hazarika D, Mazumder N, Zadeh A, Morency L-P (2017) Context-dependent sentiment analysis in user-generated videos. ACL
289. Poria S, Cambria E, Hazarika D, Mazumder N, Zadeh A, Morency L-P (2017) Multi-level multiple attentions for contextual multimodal sentiment analysis. In: IEEE International Conference on Data Mining Series (ICDM), vol 1
290. Poria S, Cambria E, Howard N, Huang G-B, Hussain A (2016) Fusing audio, visual and textual clues for sentiment analysis from multimodal content. Neurocomputing 174:50–59
291. Poria S, Cambria E, Hussain A, Huang G-B (2015) Towards an intelligent framework for multimodal affective data analysis. Neural Netw 63:104–116
292. Poria S, Cambria E, Winterstein G, Huang G-B (2014) Sentic patterns: dependency-based rules for concept-level sentiment analysis. Knowl-Based Syst 69:45–63
293. Poria S, Chaturvedi I, Cambria E, Bisio F (2016) Sentic LDA: improving on LDA with semantic similarity for aspect-based sentiment analysis. In: 2016 International Joint Conference on Neural Networks (IJCNN). IEEE, pp 4465–4473
294. See Ref. [294].
295. Poria S, Chaturvedi I, Cambria E, Hussain A (2016) Convolutional MKL based multimodal emotion recognition and sentiment analysis. In: 2016 IEEE 16th International Conference on Data Mining (ICDM). IEEE, pp 439–448
296. Poria S, Gelbukh A, Agarwal B, Cambria E, Howard N (2013) Common sense knowledge based personality recognition from text. In: Mexican International Conference on Artificial Intelligence. Springer, Berlin/Heidelberg, pp 484–496
297. Poria S, Gelbukh A, Agarwal B, Cambria E, Howard N (2014) Sentic demo: a hybrid concept-level aspect-based sentiment analysis toolkit. In: ESWC 2014
298. Poria S, Gelbukh A, Cambria E, Das D, Bandyopadhyay S (2012) Enriching senticnet polarity scores through semi-supervised fuzzy clustering. In: 2012 IEEE 12th International Conference on Data Mining Workshops (ICDMW). IEEE, pp 709–716
299. Poria S, Gelbukh A, Cambria E, Hussain A, Huang G-B (2014) Emosenticspace: a novel framework for affective common-sense reasoning. Knowl-Based Syst 69:108–123
300. Poria S, Gelbukh A, Cambria E, Yang P, Hussain A, Durrani T (2012) Merging senticnet and wordnet-affect emotion lists for sentiment analysis. In: 2012 IEEE 11th International Conference on Signal Processing (ICSP), vol 2. IEEE, pp 1251–1255
301. Poria S, Gelbukh A, Das D, Bandyopadhyay S (2012) Fuzzy clustering for semi-supervised learning—case study: construction of an emotion lexicon. In: Proceedings of MICAI
302. Poria S, Gelbukh A, Das D, Bandyopadhyay S (2013) Fuzzy clustering for semi-supervised learning–case study: construction of an emotion lexicon. In: Advances in artificial intelligence. Springer, pp 73–86
303. Poria S, Gelbukh A, Hussain A, Howard N, Das D, Bandyopadhyay S (2013) Enhanced senticnet with affective labels for concept-based opinion mining. IEEE Intell Syst 28(2): 31–38

304. Poria S, Peng H, Hussain A, Howard N, Cambria E (2017) Ensemble application of convolutional neural networks and multiple kernel learning for multimodal sentiment analysis. Neurocomputing 261:217–230
305. Potamitis I, Chen H, Tremoulis G (2004) Tracking of multiple moving speakers with multiple microphone arrays. IEEE Trans Speech Audio Process 12(5):520–529
306. Potts C (2005) The logic of conventional implicatures. Oxford University Press, Oxford
307. Prinz JJ (2004) Gut reactions: a perceptual theory of emotion. Oxford University Press, Oxford/New York
308. Pudil P, Ferri FJ, Novovicova J, Kittler J (1994) Floating search methods for feature selection with nonmonotonic criterion functions. In: Proceedings of the 12th IAPR International (1994) Conference on Pattern Recognition, vol 2-Conference B: Computer Vision and Image Processing, vol 2. IEEE, pp 279–283
309. Qi H, Wang X, Sitharama Iyengar S, Chakrabarty K (2001) Multisensor data fusion in distributed sensor networks using mobile agents. In: Proceedings of 5th International Conference on Information Fusion, pp 11–16
310. Qiu G, Liu B, Bu J, Chen C (2009) Expanding domain sentiment lexicon through double propagation. In: IJCAI, vol 9, pp 1199–1204
311. Rabie A, Wrede B, Vogt T, Hanheide M (2009) Evaluation and discussion of multimodal emotion recognition. In: Second International Conference on Computer and Electrical Engineering ICCEE'09 , vol 1. IEEE, pp 598–602
312. Rahman T, Busso C (2012) A personalized emotion recognition system using an unsupervised feature adaptation scheme. In: 2012 IEEE International Conference on Acoustics, Speech and Signal Processing (ICASSP). IEEE, pp 5117–5120
313. Rajagopal D, Cambria E, Olsher D, Kwok K (2013) A graph-based approach to commonsense concept extraction and semantic similarity detection. In: Proceedings of the 22nd International Conference on World Wide Web. ACM, pp 565–570
314. Ramesh BP, Yu H (2010) Identifying discourse connectives in biomedical text. In: AMIA Annual Symposium Proceedings, vol 2010. American Medical Informatics Association, p 657
315. Rashid M, Abu-Bakar SAR, Mokji M (2013) Human emotion recognition from videos using spatio-temporal and audio features. Vis Comput 29(12):1269–1275
316. Rauber TW, de Assis Boldt F, Varejão FM (2015) Heterogeneous feature models and feature selection applied to bearing fault diagnosis. IEEE Trans Ind Electron 62(1):637–646
317. Razavian AS, Azizpour H, Sullivan J, Carlsson S (2014) CNN features off-the-shelf: an astounding baseline for recognition. In: 2014 IEEE Conference on Computer Vision and Pattern Recognition Workshops (CVPRW). IEEE, pp 512–519
318. Repp S (2009) Negation in gapping. Oxford University Press, Oxford
319. Resnik P (1995) Using information content to evaluate semantic similarity in a taxonomy. arXiv preprint cmp-lg/9511007
320. Reyes A, Rosso P (2014) On the difficulty of automatically detecting irony: beyond a simple case of negation. Knowl Inf Syst 40(3):595–614
321. Ridella S, Rovetta S, Zunino R (1997) Circular backpropagation networks for classification. IEEE Trans Neural Netw 8(1):84–97
322. Rigoll G, Müller R, Schuller B (2005) Speech emotion recognition exploiting acoustic and linguistic information sources. In: Proceedings of SPECOM, Patras, pp 61–67
323. Rinn WE (1984) The neuropsychology of facial expression: a review of the neurological and psychological mechanisms for producing facial expressions. Psychol Bull 95(1):52
324. Rosas V, Mihalcea R, Morency L-P (2013) Multimodal sentiment analysis of Spanish online videos. IEEE Intell Syst 28(3):38–45
325. Rosasco L, De Vito E, Caponnetto A, Piana M, Verri A (2004) Are loss functions all the same? Neural Comput 16(5):1063–1076
326. Rozgic V, Ananthakrishnan S, Saleem S, Kumar R, Prasad R (2012) Speech language & multimedia technologies, Raytheon BBN technologies, Cambridge, MA. In: 2012 Asia-Pacific Signal & Information Processing Association Annual Summit and Conference (APSIPA ASC). IEEE, pp 1–4

327. Rozgić V, Ananthakrishnan S, Saleem S, Kumar R, Prasad R (2012) Ensemble of SVM trees for multimodal emotion recognition. In: 2012 Asia-Pacific Signal & Information Processing Association Annual Summit and Conference (APSIPA ASC). IEEE, pp 1–4
328. Russell JA (1979) Affective space is bipolar. J Pers Soc Psychol 37(3):345–356
329. Russell JA (2003) Core affect and the psychological construction of emotion. Psychol Rev 110(1):145
330. Sahlgren M, Cöster R (2004) Using bag-of-concepts to improve the performance of support vector machines in text categorization. In: Proceedings of the 20th International Conference on Computational Linguistics. Association for Computational Linguistics, p 487
331. Saragih JM, Lucey S, Cohn JF (2009) Face alignment through subspace constrained mean-shifts. In: 2009 IEEE 12th International Conference on Computer Vision. IEEE, pp 1034–1041
332. Sarkar C, Bhatia S, Agarwal A, Li J (2014) Feature analysis for computational personality recognition using Youtube personality data set. In: Proceedings of the 2014 ACM Multimedia on Workshop on Computational Personality Recognition. ACM, pp 11–14
333. Savran A, Cao H, Shah M, Nenkova A, Verma R (2012) Combining video, audio and lexical indicators of affect in spontaneous conversation via particle filtering. In: Proceedings of the 14th ACM International Conference on Multimodal Interaction. ACM, pp 485–492
334. Scharl A, Hubmann-Haidvogel A, Jones A, Fischl D, Kamolov R, Weichselbraun A, Rafelsberger W (2016) Analyzing the public discourse on works of fiction – detection and visualization of emotion in online coverage about HBO's game of thrones. Inf Process Manag 52(1):129–138
335. Scherer KR (1996) Adding the affective dimension: a new look in speech analysis and synthesis. In: ICSLP
336. Scherer KR (2005) What are emotions? and how can they be measured? Soc Sci Inf 44(4):695–729
337. Schuller B, Müeller R, Höernler B, Höethker A, Konosu H, Rigoll G (2007) Audiovisual recognition of spontaneous interest within conversations. In: Proceedings of the 9th International Conference on Multimodal Interfaces. ACM, pp 30–37
338. Schuller B (2011) Recognizing affect from linguistic information in 3D continuous space. IEEE Trans Affect Comput 2(4):192–205
339. Sebe N, Cohen I, Gevers T, Huang TS (2006) Emotion recognition based on joint visual and audio cues. In: 18th International Conference on Pattern Recognition ICPR 2006, vol 1. IEEE, pp 1136–1139
340. Seppi D, Batliner A, Schuller B, Steidl S, Vogt T, Wagner J, Devillers L, Vidrascu L, Amir N, Aharonson V (2008) Patterns, prototypes, performance: classifying emotional user states. In: INTERSPEECH, pp 601–604
341. Shah RR (2016) Multimodal analysis of user-generated content in support of social media applications. In: Proceedings of the ACM International Conference on Multimedia Retrieval, pp 423–426
342. Shah RR (2016) Multimodal-based multimedia analysis, retrieval, and services in support of social media applications. In: Proceedings of the ACM International Conference on Multimedia
343. Shah RR, Hefeeda M, Zimmermann R, Harras K, Hsu C-H, Yu Y (2016) NEWSMAN: uploading videos over adaptive middleboxes to news servers in weak network infrastructures. In: Proceedings of the Springer International Conference on Multimedia Modeling, pp 100–113
344. Shah RR, Samanta A, Gupta D, Yu Y, Tang S, Zimmermann R (2016) PROMPT: personalized user tag recommendation for social media photos leveraging multimodal information. In: Proceedings of the ACM International Conference on Multimedia, pp 486–492
345. Shah RR, Shaikh AD, Yu Y, Geng W, Zimmermann R, Wu G (2015) EventBuilder: real-time multimedia event summarization by visualizing social media. In: Proceedings of the ACM International Conference on Multimedia, pp 185–188

346. Shah RR, Yu Y, Shaikh AD, Tang S, Zimmermann R (2014) ATLAS: automatic temporal segmentation and annotation of lecture videos based on modelling transition time. In: Proceedings of the ACM International Conference on Multimedia, pp 209–212
347. Shah RR, Yu Y, Shaikh AD, Zimmermann R (2015) TRACE: a linguistic-based approach for automatic lecture video segmentation leveraging Wikipedia texts. In: Proceedings of the IEEE International Symposium on Multimedia, pp 217–220
348. Shah RR, Yu Y, Tang S, Satoh S, Verma A, Zimmermann R (2016) Concept-level multimodal ranking of Flickr photo tags via recall based weighting. In: Proceedings of the MMCommon's Workshop at ACM International Conference on Multimedia, pp 19–26
349. Shah RR, Yu Y, Verma A, Tang S, Shaikh A, Zimmermann R (2016) Leveraging multimodal information for event summarization and concept-level sentiment analysis. Knowl-Based Syst 108:102–109
350. Shah RR, Yu Y, Zimmermann R (2014) ADVISOR: personalized video soundtrack recommendation by late fusion with heuristic rankings. In: Proceedings of the ACM International Conference on Multimedia, pp 607–616
351. Shah RR, Yu Y, Zimmermann R (2014) User preference-aware music video generation based on modeling scene moods. In: Proceedings of the ACM International Conference on Multimedia Systems, pp 156–159
352. Shaikh AD, Jain M, Rawat M, Shah RR, Kumar M (2013) Improving accuracy of SMS based FAQ retrieval system. In: Proceedings of the Springer Multilingual Information Access in South Asian Languages, pp 142–156
353. Shaikh AD, Shah RR, Shaikh R (2013) SMS based FAQ retrieval for Hindi, English and Malayalam. In: Proceedings of the ACM Forum on Information Retrieval Evaluation, p 9
354. Shan C, Gong S, McOwan PW (2007) Beyond facial expressions: learning human emotion from body gestures. In: BMVC, pp 1–10
355. Shimojo S, Shams L (2001) Sensory modalities are not separate modalities: plasticity and interactions. Curr Opin Neurobiol 11(4):505–509
356. Siddiquie B, Chisholm D, Divakaran A (2015) Exploiting multimodal affect and semantics to identify politically persuasive web videos. In: Proceedings of the 2015 ACM on International Conference on Multimodal Interaction. ACM, pp 203–210
357. Sidorov G, Posadas-Durán J-P, Salazar HJ, Chanona-Hernandez L (2011) A new combined lexical and statistical based sentence level alignment algorithm for parallel texts. Int J Comput Ling Appl 2(1–2):257–263
358. Simons M, Tonhauser J, Beaver D, Roberts C (2010) What projects and why. In: Proceedings of Semantics and Linguistic Theory (SALT) 20, pp 309–327
359. Socher R, Huval B, Manning CD, Ng AY (2012) Semantic compositionality through recursive matrix-vector spaces. In: Proceedings of the 2012 Joint Conference on Empirical Methods in Natural Language Processing and Computational Natural Language Learning, Association for Computational Linguistics, pp 1201–1211
360. Socher R, Perelygin A, Wu JY, Chuang J, Manning CD, Ng AY, Potts C (2013) Recursive deep models for semantic compositionality over a sentiment treebank. In: Proceedings of the Conference on Empirical Methods in Natural Language Processing (EMNLP), vol 1631. Citeseer, p 1642
361. See Ref. [359].
362. Soleymani M, Pantic M, Pun T (2012) Multimodal emotion recognition in response to videos. IEEE Trans Affect Comput 3(2):211–223
363. Song M, Bu J, Chen C, Li N (2004) Audio-visual based emotion recognition-a new approach. In: Computer vision and pattern recognition, vol 2. IEEE, pp II–1020
364. Speer R, Havasi C (2013) Conceptnet 5: a large semantic network for relational knowledge. In: The People's web meets NLP. Springer, pp 161–176
365. See Ref. [365].
366. Strapparava C, Valitutti A et al (2004) Wordnet affect: an affective extension of wordnet. In: LREC, vol 4, pp 1083–1086
367. Subrahmanya N, Shin YC (2010) Sparse multiple kernel learning for signal processing applications. IEEE Trans Pattern Anal Mach Intell 32(5):788–798

368. Tajadura-Jiménez A, Västfjäll D (2008) Auditory-induced emotion: a neglected channel for communication in human-computer interaction. In: Affect and emotion in human-computer interaction. Springer, pp 63–74
369. Tao H, Chen HH, Wu W, Huang TS (1999) Compression of MPEG-4 facial animation parameters for transmission of talking heads. IEEE Trans Circuits Syst Video Technol 9(2):264–276
370. Tian Y-L, Kanade T, Cohn JF (2001) Recognizing action units for facial expression analysis. IEEE Trans Pattern Anal Mach Intell 23(2):97–115
371. Tran D, Bourdev L, Fergus R, Torresani L, Paluri M (2014) Learning spatiotemporal features with 3D convolutional networks. arXiv preprint arXiv:1412.0767
372. Tu B, Yu F (2012) Bimodal emotion recognition based on speech signals and facial expression. In: Foundations of intelligent systems. Springer, Berlin, pp 691–696
373. Turney PD (2002) Thumbs up or thumbs down? semantic orientation applied to unsupervised classification of reviews. In: Proceedings of the 40th Annual Meeting on Association for Computational Linguistics. Association for Computational Linguistics, pp 417–424
374. Valstar MF, Almaev T, Girard JM, McKeown G, Mehu M, Yin L, Pantic M, Cohn JF (2015) Fera 2015-second facial expression recognition and analysis challenge. In: Automatic face and gesture recognition, vol 6, pp 1–8
375. Västfjäll D, Kleiner M (2002) Emotion in product sound design. In: Proceedings of Journées Design Sonore
376. Verma R, Davatzikos C, Loughead J, Indersmitten T, Hu R, Kohler C, Gur RE, Gur RC (2005) Quantification of facial expressions using high-dimensional shape transformations. J Neurosci Methods 141(1):61–73
377. Vicente L (2010) On the syntax of adversative coordination. Nat Lang Linguist Theory 28(2):381–415
378. Vogl TP, Mangis JK, Rigler AK, Zink WT, Alkon DL (1988) Accelerating the convergence of the back-propagation method. Biol Cybern 59(4–5):257–263
379. Vogt T, André E (2005) Comparing feature sets for acted and spontaneous speech in view of automatic emotion recognition. In: IEEE International Conference on Multimedia and Expo ICME 2005. IEEE, pp 474–477
380. Vogt T, André E, Wagner J (2008) Automatic recognition of emotions from speech: a review of the literature and recommendations for practical realisation. In: Affect and emotion in human-computer interaction. Springer, pp 75–91
381. Vu HA, Yamazaki Y, Dong F, Hirota K (2011) Emotion recognition based on human gesture and speech information using RT middleware. In: 2011 IEEE International Conference on Fuzzy Systems (FUZZ). IEEE, pp 787–791
382. Wagner J, Andre E, Lingenfelser F, Kim J (2011) Exploring fusion methods for multimodal emotion recognition with missing data. IEEE Trans Affect Comput 2(4):206–218
383. Walter S, Scherer S, Schels M, Glodek M, Hrabal D, Schmidt M, Böck R, Limbrecht K, Traue HC, Schwenker F (2011) Multimodal emotion classification in naturalistic user behavior. In: Human-computer interaction. Towards mobile and intelligent interaction environments. Springer, pp 603–611
384. Wang F, Wang Z, Li Z, Wen J-R (2014) Concept-based short text classification and ranking. In: Proceedings of the 23rd ACM International Conference on Information and Knowledge Management. ACM, pp 1069–1078
385. Wang S, Zhu Y, Wu G, Ji Q (2014) Hybrid video emotional tagging using users' EEG and video content. Multimedia Tools Appl 72(2):1257–1283
386. Wang Y, Guan L (2008) Recognizing human emotional state from audiovisual signals*. IEEE Trans Multimedia 10(5):936–946
387. Wang Y, Ai H, Wu B, Huang C (2004) Real time facial expression recognition with adaboost. In: Proceedings of the 17th International Conference on, vol 3. IEEE, pp 926–929
388. Wawer A (2012) Extracting emotive patterns for languages with rich morphology. Int J Comput Ling Appl 3(1):11–24

389. Wellner B, Pustejovsky J, Havasi C, Rumshisky A, Sauri R (2009) Classification of discourse coherence relations: an exploratory study using multiple knowledge sources. In: Proceedings of the 7th SIGdial Workshop on Discourse and Dialogue. Association for Computational Linguistics, pp 117–125
390. Wen Z, Huang TS (2003) Capturing subtle facial motions in 3D face tracking. In: Ninth IEEE International Conference on Computer Vision Proceedings. IEEE, pp 1343–1350
391. Whissell C (1989) The dictionary of affect in language. Emot Theory Res Exp 4(113–131):94
392. Wilson T, Wiebe J, Hoffmann P (2005) Recognizing contextual polarity in phrase-level sentiment analysis. In: Proceedings of the Conference on Human Language Technology and Empirical Methods in Natural Language Processing. Association for Computational Linguistics, pp 347–354
393. Wimmer M, Schuller B, Arsic D, Rigoll G, Radig B (2008) Low-level fusion of audio, video feature for multi-modal emotion recognition. In: VISAPP, vol 2, pp 145–151
394. Winterstein G (2012) What but-sentences argue for: a modern argumentative analysis of but. Lingua 122(15):1864–1885
395. Wolf F, Gibson E (2005) Representing discourse coherence: a corpus-based study. Comput Ling 31(2):249–287
396. Wöllmer M, Kaiser M, Eyben F, Schuller B, Rigoll G (2013) LSTM-modeling of continuous emotions in an audiovisual affect recognition framework. Image Vis Comput 31(2):153–163
397. Wöllmer M, Metallinou A, Eyben F, Schuller B, Narayanan SS et al (2010) Context-sensitive multimodal emotion recognition from speech and facial expression using bidirectional LSTM modeling. In: Interspeech, pp 2362–2365
398. Wollmer M, Weninger F, Knaup T, Schuller B, Sun C, Sagae K, Morency L-P (2013) Youtube movie reviews: sentiment analysis in an audio-visual context. IEEE Intell Syst 28(3):46–53
399. Wu C-E, Tsai RT-H (2014) Using relation selection to improve value propagation in a conceptnet based sentiment dictionary. Knowl-Based Syst 69:100–107
400. Wu C-H, Liang W-B (2011) Emotion recognition of affective speech based on multiple classifiers using acoustic-prosodic information and semantic labels. IEEE Trans Affect Comput 2(1):10–21
401. Wu C-H, Yeh J-F, Chuang Z-J (2009) Emotion perception and recognition from speech. In: Tao J, Tan T (eds) Affective information processing. Springer, London, pp 93–110
402. Wu H-H, Charng-Rurng Tsai A, Tzong-Han Tsai R, Yung-jen Hsu J (2011) Sentiment value propagation for an integral sentiment dictionary based on commonsense knowledge. In: 2011 International Conference on Technologies and Applications of Artificial Intelligence (TAAI). IEEE, pp 75–81
403. Wu Z, Palmer M (1994) Verbs semantics and lexical selection. In: Proceedings of the 32nd Annual Meeting on Association for Computational Linguistics. Association for Computational Linguistics, pp 133–138
404. Xie H, Li X, Wang T, Lau R, Wong T-L, Chen L, Wang F-L, Li Q (2016) Incorporating sentiment into tag-based user profiles and resource profiles for personalized search in folksonomy. Inf Process Manag 52:61–72
405. Xu B, Fu Y, Jiang Y-G, Li B, Sigal L (2015) Heterogeneous knowledge transfer in video emotion recognition, attribution and summarization. arXiv preprint arXiv:1511.04798
406. Xu C, Cetintas S, Lee K-C, Li L-J (2014) Visual sentiment prediction with deep convolutional neural networks. arXiv preprint arXiv:1411.5731
407. Yacoob Y, Davis L (1994) Computing spatio-temporal representations of human faces. In: Computer vision and pattern recognition. IEEE, pp 70–75
408. Yamasaki T, Fukushima Y, Furuta R, Sun L, Aizawa K, Bollegala D (2015) Prediction of user ratings of oral presentations using label relations. In: Proceedings of the 1st International Workshop on Affect & Sentiment in Multimedia. ACM, pp 33–38
409. Yang C, Hsin-Yih Lin K, Chen H-H (2007) Building emotion lexicon from weblog corpora. In: Proceedings of the 45th Annual Meeting of the ACL on Interactive Poster and Demonstration Sessions. Association for Computational Linguistics, pp 133–136

410. Yeasin M, Bullot B, Sharma R (2004) From facial expression to level of interest: a spatio-temporal approach. In: Computer vision and pattern recognition, vol 2. IEEE, pp II–922
411. Yin Y, Shah RR, Zimmermann R (2016) A general feature-based map matching framework with trajectory simplification. In: Proceedings of the 7th ACM SIGSPATIAL International Workshop on GeoStreaming, p 7
412. You Q, Luo J, Jin H, Yang J (2015) Robust image sentiment analysis using progressively trained and domain transferred deep networks. arXiv preprint arXiv:1509.06041
413. Young T, Hazarika D, Poria S, Cambria E (2017) Recent trends in deep learning based natural language processing. arXiv preprint arXiv:1708.02709
414. Yu H, Hatzivassiloglou V (2003) Towards answering opinion questions: separating facts from opinions and identifying the polarity of opinion sentences. In: Proceedings of the 2003 Conference on Empirical Methods in Natural Language Processing. Association for Computational Linguistics, pp 129–136
415. Yu H, Yang J, Han J (2003) Classifying large data sets using SVMS with hierarchical clusters. In: Proceedings of the Ninth ACM SIGKDD International Conference on Knowledge Discovery and Data Mining. ACM, pp 306–315
416. Zadeh A, Chen M, Poria S, Cambria E, Morency L-P (2017) Tensor fusion network for multimodal sentiment analysis. In: EMNLP. EMNLP
417. Zadeh A, Liang PP, Mazumder N, Poria S, Cambria E, Morency L-P (2018) Memory fusion network for multi-view sequential learning. arXiv preprint arXiv:1802.00927
418. Zadeh A, Liang PP, Poria S, Vij P, Cambria E, Morency L-P (2018) Multi-attention recurrent network for human communication comprehension. arXiv preprint arXiv:1802.00923
419. Zeng Z, Hu Y, Fu Y, Huang TS, Roisman GI, Wen Z (2006) Audio-visual emotion recognition in adult attachment interview. In: Proceedings of the 8th International Conference on Multimodal Interfaces. ACM, pp 139–145
420. Zeng Z, Hu Y, Liu M, Fu Y, Huang TS (2006) Training combination strategy of multi-stream fused hidden Markov model for audio-visual affect recognition. In: Proceedings of the 14th Annual ACM International Conference on Multimedia. ACM, pp 65–68
421. Zeng Z, Pantic M, Roisman GI, Huang TS (2009) A survey of affect recognition methods: audio, visual, and spontaneous expressions. IEEE Trans Pattern Anal Mach Intell 31(1):39–58
422. Zeng Z, Tu J, Liu M, Huang TS (2005) Multi-stream confidence analysis for audio-visual affect recognition. In: Affective computing and intelligent interaction. Springer, pp 964–971
423. Zeng Z, Tu J, Liu M, Huang TS, Pianfetti B, Roth D, Levinson S (2007) Audio-visual affect recognition. IEEE Trans Multimedia 9(2):424–428
424. Zhang S, Li L, Zhao Z (2012) Audio-visual emotion recognition based on facial expression and affective speech. In: Multimedia and signal processing. Springer, pp 46–52
425. Zhang T, Yu B et al (2005) Boosting with early stopping: convergence and consistency. Ann Stat 33(4):1538–1579
426. Zhang Y, Liu B (2007) Semantic text classification of emergent disease reports. In: Knowledge discovery in databases: PKDD 2007. Springer, Berlin/Heidelberg pp 629–637
427. Zhang Z (1999) Feature-based facial expression recognition: sensitivity analysis and experiments with a multilayer perceptron. Int J Pattern Recogn Artif Intell 13(06):893–911
428. Zuo Z, Shuai B, Wang G, Liu X, Wang X, Wang B, Chen Y (2015) Convolutional recurrent neural networks: learning spatial dependencies for image representation. In: 2015 IEEE Conference on Computer Vision and Pattern Recognition Workshops, CVPR Workshops, Boston, 7–12 June 2015, pp 18–26

Index

S. Poria et al., *Multimodal Sentiment Analysis*, Socio-Affective Computing 8,
https://doi.org/10.1007/978-3-319-95020-4

T

U

V

MIX
Papier aus verantwortungsvollen Quellen
Paper from responsible sources
FSC® C105338

If you have any concerns about our products,
you can contact us on
ProductSafety@springernature.com

In case Publisher is established outside the EU,
the EU authorized representative is:
Springer Nature Customer Service Center GmbH
Europaplatz 3, 69115 Heidelberg, Germany

Printed by Libri Plureos GmbH
in Hamburg, Germany